SpringerBriefs in Computer Science

SpringerBriefs present concise summaries of cutting-edge research and practical applications across a wide spectrum of fields. Featuring compact volumes of 50 to 125 pages, the series covers a range of content from professional to academic.

Typical topics might include:

- A timely report of state-of-the art analytical techniques
- A bridge between new research results, as published in journal articles, and a contextual literature review
- A snapshot of a hot or emerging topic
- An in-depth case study or clinical example
- A presentation of core concepts that students must understand in order to make independent contributions

Briefs allow authors to present their ideas and readers to absorb them with minimal time investment. Briefs will be published as part of Springer's eBook collection, with millions of users worldwide. In addition, Briefs will be available for individual print and electronic purchase. Briefs are characterized by fast, global electronic dissemination, standard publishing contracts, easy-to-use manuscript preparation and formatting guidelines, and expedited production schedules. We aim for publication 8–12 weeks after acceptance. Both solicited and unsolicited manuscripts are considered for publication in this series.

**Indexing: This series is indexed in Scopus, Ei-Compendex, and zbMATH **

Vipin Tyagi

Digital Image Forgery Detection

Techniques, Challenges and Applications

Vipin Tyagi
Computer Science and Engineering
Jaypee University of Engineering and Technology
Guna, Madhya Pradesh, India

ISSN 2191-5768　　　　　　　ISSN 2191-5776　(electronic)
SpringerBriefs in Computer Science
ISBN 978-981-95-3003-8　　　ISBN 978-981-95-3004-5　(eBook)
https://doi.org/10.1007/978-981-95-3004-5

© The Editor(s) (if applicable) and The Author(s), under exclusive license to Springer Nature Singapore Pte Ltd. 2026

This work is subject to copyright. All rights are solely and exclusively licensed by the Publisher, whether the whole or part of the material is concerned, specifically the rights of translation, reprinting, reuse of illustrations, recitation, broadcasting, reproduction on microfilms or in any other physical way, and transmission or information storage and retrieval, electronic adaptation, computer software, or by similar or dissimilar methodology now known or hereafter developed.
The use of general descriptive names, registered names, trademarks, service marks, etc. in this publication does not imply, even in the absence of a specific statement, that such names are exempt from the relevant protective laws and regulations and therefore free for general use.
The publisher, the authors and the editors are safe to assume that the advice and information in this book are believed to be true and accurate at the date of publication. Neither the publisher nor the authors or the editors give a warranty, expressed or implied, with respect to the material contained herein or for any errors or omissions that may have been made. The publisher remains neutral with regard to jurisdictional claims in published maps and institutional affiliations.

This Springer imprint is published by the registered company Springer Nature Singapore Pte Ltd.
The registered company address is: 152 Beach Road, #21-01/04 Gateway East, Singapore 189721, Singapore

If disposing of this product, please recycle the paper.

Foreword

Since the early days of photography, images have carried the aura of authenticity. They have been regarded as objective witnesses of reality, and their evidentiary value has been accepted almost without hesitation in domains as diverse as journalism, scientific research, law, and history. Yet, as we are all aware, this perception has never been entirely justified. Even in the analog era, photographs could be staged, retouched, or composited. With the advent of digital technologies, however, manipulation has reached a new level of sophistication, scale, and accessibility. Today, forged images can be created and disseminated with unprecedented ease, under mining our collective confidence in the visual record.

This book *Digital Image Forgery Detection* addresses this challenge with both breadth and depth. It begins by situating the problem of image forgery within a historical continuum, reminding us that deception by visual means is not new, though its digital manifestations are qualitatively different. The author then proceeds to develop a taxonomy of forgery types, ranging from subtle modifications of genuine photographs to the generation of entirely synthetic images by computer. This classification provides a valuable framework for understanding the diversity of threats we face.

The technical core of the book is a systematic and comprehensive survey of forgery detection methodologies. Traditional approaches based on pixel-level statistics, image format artifacts, camera characteristics, and geometric consistency are explained with clarity. These are then complemented by a presentation of modern machine learning approaches, particularly those grounded in deep neural networks. The coverage of convolutional, recurrent, adversarial, and transformer-based models, as well as hybrid and localization-focused architectures, illustrates the remarkable progress that has been achieved in recent years. At the same time, the authors do not overlook the practical difficulties of constructing reliable systems and devote well-deserved attention to datasets, evaluation metrics, and the challenges of generalization and robustness.

The dedication and clarity with which the author has brought all of these elements together is very impressive. The author of this book has done an excellent job in synthesizing a vast body of knowledge, presenting it with both rigor and accessibility. It is not a simple survey, but a carefully curated narrative that reflects deep understanding of the subject, sensitivity to its challenges, and an admirable commitment to advancing the field. One can clearly see the author's effort to bridge the gap between theoretical foundations, practical implementations, and societal implications. Such a contribution is not only valuable but also necessary at a time when trust in digital imagery is increasingly under threat.

In bringing together historical background, methodological detail, practical challenges, and future directions, this book constitutes a significant contribution to the growing literature on digital image forensics. It will be of great value to those entering the field as well as to established researchers seeking a comprehensive reference.

I warmly commend this book to its readers. It provides not only an authoritative account of the state of the art, but also a clear indication of the path forward in our collective effort to secure trust in visual media.

Institute of Information Theory and Automation   Jan Flusser
Czech Academy of Sciences,
Prague, Czechia

Preface

In the rapidly evolving digital landscape, images play a crucial role in communication, journalism, entertainment, and even legal evidence. Digital images are used to convey information and shape perceptions, making their authenticity and integrity paramount. However, the ease with which sophisticated image editing tools are now available poses a significant threat, eroding trust and creating the need for robust image forgery detection methods. In the current era of prevalent misinformation, distinguishing between authentic and digitally altered photographs is a difficult task. This situation undermines the trustworthiness and authenticity of content, making it challenging to ensure reliability in an increasingly visually driven world.

The prevalence of fake news, propaganda, and malicious use of manipulated images necessitates the development and understanding of efficient and reliable forgery detection methods. As image manipulation continues to grow in sophistication, so too must detection techniques to detect and defend against it.

This book, *Digital Image Forgery Detection: Techniques, Challenges and Applications*, presents a comprehensive exploration of the theories, algorithms, tools, and technologies developed to detect and localize image manipulations. It provides a structured examination of both traditional approaches such as active methods using watermarks and signatures and modern passive techniques, including pixel-based, statistical, and deep learning models. Special attention is given to emerging areas such as GAN-generated image detection, multimodal Deepfake detection, and transformer-based architectures, which represent the cutting edge of the field. This book will equip readers with the knowledge and tools to identify and analyze forged images, contributing to a more informed and secure digital environment. A bibliography of the literature in the area has also been provided for the help of researchers.

This work is the culmination of years of academics and research in the domain of digital image forensics. Inputs of my scholars involved in research on digital image forgery detection, especially Dr. K. B. Meena is thankfully acknowledged.

I wish to express my heartfelt gratitude to my teacher Prof. V. K. Agarwal and my mother, Smt. Sarla Tyagi, whose blessings have always guided me; to my wife, Ms. Meenakshi, for her constant encouragement and to my children, Kriti Tyagi and

Nikunj Tyagi, whose love and patience have been a source of strength throughout this journey.

Targeted at technical students, researchers, forensic analysts, and professionals in the fields of image forensics, cybersecurity, and computer vision, this SpringerBriefs will serve as a guide to one of the most important problems in the digital age.

My sincere thanks to all those who have contributed directly or indirectly in the completion of the book. The research work of several researchers has contributed in preparation of the book, I thankfully acknowledge their work.

It has been a pleasure working with Springer Nature in the development of the book. Thanks go to all authorities involved in publishing the book and for handling the publication.

Guna, India   Vipin Tyagi

Competing Interests The author has no competing interests to declare that are relevant to the content of this manuscript.

Contents

1	**Introduction to Image Forgery**		1
	1.1	Introduction	1
	1.2	History of Photography	2
		1.2.1 Early Optical Devices (Before 1800)	2
		1.2.2 The First Photograph (1826–1839)	2
		1.2.3 Technological Advancements (1840s–1880s)	2
		1.2.4 Photography with Kodak Camera (1888–1930s)	3
		1.2.5 Emergence of Color Photography (1900s–1950s)	3
		1.2.6 Digital Revolution (1980s–2000s)	4
		1.2.7 Photography in the Smartphone Era (2010–Present)	4
	1.3	Image Forgery	4
	1.4	Conclusion	8
	References		8
2	**Types of Digital Image Forgery**		11
	2.1	Introduction	11
	2.2	Manipulation in Photographic Digital Images	12
		2.2.1 Legitimate Image Processing	12
		2.2.2 Illegitimate Image Processing or Image Forgery	13
	2.3	Computer Generated Images	17
	2.4	Conclusion	18
	References		18
3	**Digital Image Forgery Detection Techniques**		21
	3.1	Introduction	21
	3.2	Image Forgery Detection Techniques	22
	3.3	Active Image Forgery Detection Techniques	23
	3.4	Passive Image Forgery Detection Techniques	24
		3.4.1 Traditional Image Forgery Detection Techniques	24
		3.4.2 Traditional Computer Generated (Synthetic) Image Detection Techniques	27

        3.4.3   Deep Learning Based Image Forgery Detection Techniques.................................. 29
  3.5 Conclusion... 32
  References.. 32

4 Datasets and Evaluation
  4.1 Introduction... 35
  4.2 Datasets Used for Training and Testing of Image Forgery Detection Techniques....................................... 36
        4.2.1   Copy–Move Forgery Detection Related Datasets 36
        4.2.2   Image Splicing Forgery Detection Related Datasets....... 38
        4.2.3   Datasets to Evaluate an Algorithm for Photographic (Real) Image and Computer Generated (Synthetic) Image.... 39
  4.3 Evaluation of Image Forgery Detection Techniques 41
        4.3.1   Pixel-Level Metrics................................... 41
        4.3.2   Image-Level Metrics.................................. 42
        4.3.3   Region-Level Metrics................................. 42
  4.4 Conclusion... 43
  References.. 43

5 Challenges in Digital Image Forgery Detection
  5.1 Introduction... 45
  5.2 Challenges Faced in Development of an Effective Robust Image Forgery Detection Model 46
        5.2.1   Generalization of Image Forgery Detection Techniques.... 46
        5.2.2   Low Quality Images and Post-processing Operations...... 46
        5.2.3   Feature Representation and Invariance 47
        5.2.4   Image Forgery Detection Dataset Limitations............. 48
        5.2.5   Precision in Localization of Tampered Regions 49
        5.2.6   Absence of Standardized Evaluation Protocols........... 50
        5.2.7   Legal and Ethical Issues............................... 50
  5.3 Conclusion... 51
  References.. 51

6 Applications of Digital Image Forgery Detection
  6.1 Introduction... 53
  6.2 Applications in Journalism and Media 54
  6.3 Applications in Law Enforcement and Legal Frameworks 54
  6.4 Applications in Social Media and Content Moderation........... 55
  6.5 Applications in Research.................................... 56
  6.6 Conclusion... 57

7 Future Directions in Digital Image Forgery Detection Research
  7.1 Introduction... 59
  7.2 Robust and Explainable Deep Learning Models 59
  7.3 Multimodal and Cross-Modal Forgery Detection 60
  7.4 Self-Supervised and Few-Shot Learning 60

7.5	Generalization to Unknown or Emerging Forgeries	61
7.6	Spatiotemporal Forgery Localization.	61
7.7	Dataset Expansion and Realistic Benchmarking	61
7.8	Federated and Privacy-Preserving Forensics	62
7.9	Real-Time Deployment	62
7.10	Integration with Other Digital Forensics Disciplines	62
7.11	Legal and Ethical Considerations	63
7.12	Increasing Public Awareness and Education	63
7.13	Conclusion	63
	References	64

Bibliography ... 65

List of Figures

Fig. 1.1	Nicéphore Niépce's "point de vue" heliograph, the oldest surviving photograph (https://en.wikipedia.org/wiki/View_from_the_Window_at_Le_Gras)	3
Fig. 1.2	Hippolyte Bayard's self-portrait (1840) (taken from https://en.wikipedia.org/wiki/Hippolyte_Bayard)	5
Fig. 1.3	Stalin with and without Nikolai Yezhov, the chief of the NKVD, (1930s), Moscow-Volga Canal, Russian SFSR (taken from https://www.comradegallery.com/journal/fabrication-photographs-stalin-soviet-state?srsltid=AfmBOoo4aLgN4iYdN4c8fdADXrQhaHnBPgtrs3-72X6oufIsUmsnkRJv)	6
Fig. 1.4	Digitally altered photograph of the aftermath of an IDF attack on Beirut. (Reuters) https://en.wikipedia.org/wiki/File:Beirut-smoke.jpg	7
Fig. 2.1	Example of copy–move image forgery (CoMoFod database) (taken from https://www.vcl.fer.hr/comofod/examples.html)	14
Fig. 2.2	Example of copy–move image forgery with post-processing (CoMoFod database) (taken from https://www.vcl.fer.hr/comofod/postprocessing.html)	15
Fig. 2.3	Example of image splicing forgery [3]	16
Fig. 2.4	Example of image inpainting forgery (**a**) Original image (**b**) Forged image [4]	16
Fig. 2.5	Examples of computer-generated images (Top row) and photographic images (Bottom row) taken from the DSTok dataset [5]	17
Fig. 3.1	Classification of digital image forgery detection techniques	22
Fig. 3.2	A generalized framework of digital image forgery detection techniques	31

xv

About the Author

Vipin Tyagi (Prof.) is a distinguished academician, researcher, and author, currently serving as Dean (Academic and Research) at Jaypee University of Engineering and Technology in Guna, Madhya Pradesh, India. His expertise spans across Image Processing and Cyber Forensics.

Prof. Tyagi has a prolific publication record, contributing significantly to his fields of interest. His scholarly work includes numerous journal articles, conference papers, and books. Some of his notable publications are in the areas of image forgery detection, image denoising, and content-based image retrieval. He has authored and edited several books and papers in the field of computing and data sciences, such as *Predictive Computing and Information Security* and *Content-Based Image Retrieval: Ideas, Influences, and Current Trends* published by Springer Singapore, *Understanding Digital Image Processing* published by CRC Press, Taylor and Francis Group.

His scholarly work has earned him a respectable h-index and numerous citations in the academic community.

In addition to his academic pursuits, Prof. Tyagi has been actively involved in professional societies. Prof. Tyagi is a senior life member of the Computer Society of India, a Fellow of the IETE, Senior Member—IEEE, and a member of the Indian Science Congress Association, ACM. He was elected as President and Recorder of Engineering Science Section of Indian Science Congress Association. He has also served as Hon. National Secretary of the Computer Society of India. Prof. Tyagi is elected as member Board of Governors of Engineering Council of India, third time consecutively (2020–2022, 2022–2024, 2024–26).

Chapter 1
Introduction to Image Forgery

Abstract Image manipulation has evolved from physical alterations in darkrooms to sophisticated digital forgeries driven by artificial intelligence. This chapter traces the historical development of image forgery techniques, highlighting key events, and technological transitions. It covers analog manipulation, digital image editing, internet-era misinformation, and modern AI-based synthesis, culminating in forensic challenges and responses.

Keywords Image forgery · Image tempering · Forgery detection · Image forgery history · Synthetic images · Deepfake

1.1 Introduction

It is said, "A picture is worth a thousand words". This means a picture or an image may convey an information more effectively than a thousand written words. It shows the power of visual communication through an image. But if the content of the image is modified, it may communicate wrong information and convey a false message more powerfully than manipulated text ever could. This makes image forgery very dangerous. Use of digital images is enormous in our daily life to communicate information. Digital images play significant roles in advertising, animation films, computer-aided design, computational biology, simulation systems, crime scene reconstruction, web designing, virtual reality, medical, education system, online insurance claims, digital magazines, newspaper industry and to convey research findings. In all these areas, the use of digital images has brought exceptional positive changes. The authenticity of an image is an essential requirement in most of these areas. When the content of a digital image is modified or altered to change its meaning, context or appearance or the whole image is generated by a computer graphics software for wrong intention, such altered image is commonly termed as a forged image and the process of creating such forged image is known as digital image forgery. Forged images can be weapons of misinformation that are harder to detect due to their apparent authenticity.

© The Author(s), under exclusive license to Springer Nature Singapore Pte Ltd. 2026
V. Tyagi, *Digital Image Forgery Detection*, SpringerBriefs in Computer Science, https://doi.org/10.1007/978-981-95-3004-5_1

In this chapter, the evolution of photography and image forgery from its earliest photographic instances in the nineteenth century to the complex, digital and AI-powered forgery of the twenty-first century is discussed.

1.2 History of Photography

Photography, the art and science of capturing images using light, has undergone transformative developments since its inception. Evolution of photography from analog to digital has not only revolutionized communication and documentation but also enabled sophisticated manipulation, raising significant concerns in areas like research, social media posts, journalism, legal evidence, and forensic analysis. A concise summary of this journey of photography is provided below.

1.2.1 Early Optical Devices (Before 1800)

The foundation of photography lies in the principles of optics. Devices like the camera obscura, first described by Chinese philosopher Mozi (~fifth century BCE) and later by Aristotle and Alhazen, projected an inverted image of a scene onto a surface through a small aperture [1]. This device was useful for artists and scientists but there was no provision of storing the image.

1.2.2 The First Photograph (1826–1839)

The first permanent photographic image (Fig. 1.1) was created in 1826 by Joseph Nicéphore Niépceusing a process called heliography, which required several hours of exposure on a bitumen-coated pewter plate [2]. Niépce later collaborated with Louis Daguerre, who introduced the daguerreotype in 1839; a silver-coated copper plate sensitized with iodine vapor, exposed to light, and developed using mercury vapor [3]. Meanwhile, William Henry Fox Talbot developed the calotype, a paper-based negative–positive process allowing multiple copies [4].

1.2.3 Technological Advancements (1840s–1880s)

Progress in chemical processes led to improved image quality and portability. Wet Collodion Process (Frederick Scott Archer, 1851) replaced earlier techniques with glass plates coated in collodion, producing sharper images and shorter exposures

1.2 History of Photography

Fig. 1.1 Nicéphore Niépce's "point de vue" heliograph, the oldest surviving photograph (https://en.wikipedia.org/wiki/View_from_the_Window_at_Le_Gras)

[5]. Dry Plates (Richard Leach Maddox, 1871) enabled greater flexibility and ease of use, allowing photographers to develop images after exposure [6].

1.2.4 Photography with Kodak Camera (1888–1930s)

Revolution in the area of photography came with the invent of Kodak camera in 1888 by George Eastman, making it accessible to the masses. Its roll film and slogan "You press the button, we do the rest" made image capturing a popular activity [7].

1.2.5 Emergence of Color Photography (1900s–1950s)

Color photography, initially experimental, matured with the Autochromeplate (1907) developed by the Lumière brothers, followed by the Kodachromefilm (1935) introduced by Kodak, enabling vibrant color reproduction and becoming a favorite for professional photography [8]. The invention of the Polaroid instant camera by Edwin Land in 1947 allowed users to develop images instantly [9]. During this period, 35 mm film cameras and SLRs became the norm for both professional and personal photography.

1.2.6 Digital Revolution (1980s–2000s)

Digital photography emerged with the development of CCD sensors by Bell Labs and the first digital camera prototype by Kodak engineer Steve Sasson in 1975 [10]. In the 1990s, digital cameras became commercially available, leading to the decline of film-based photography.

1.2.7 Photography in the Smartphone Era (2010–Present)

Modern smartphones feature high-resolution digital cameras with advanced software capabilities like HDR, AI enhancements, and computational photography. Platforms such as WhatsApp Instagram and Facebook have further accelerated the daily production and dissemination of images [11].

The evolution of photography from analog to digital has profoundly affected the ease of manipulation. In earlier eras, image tampering required darkroom expertise; today, software like Photoshop, GANs, and deepfake generators allow realistic image forgeries with minimal effort, challenging the authenticity of visual content in media, legal, and academic contexts.

1.3 Image Forgery

Any alteration in an image with the bad intention is known as image forgery. If this forgery is performed on a digital image then image forgery is called digital image forgery. Digital image forgery types are discussed in detail in Chap. 2 of this book.

The history of image manipulation began soon after the invention of photography in the nineteenth century. Hippolyte Bayard, assumed to be the first person to create a fake image (Fig. 1.2) as recorded in history, is famous for a picture of him committing suicide. This image is Hippolyte Bayard's self-portrait (1840), where he depicted himself as a drowned man, a staged protest against the neglect of his photographic invention [12]. It was later revealed that this photograph was forged due to his frustration because he had lost the chance of becoming 'the inventor of photography' to Louis Daguerre. Daguerre patented a photography process earlier than him and owned all the fame [13].

Early photographers used physical techniques in analog form to manipulate images such as: Scratching or painting negatives, combining multiple negatives (composite photography) to generate an image.

By the early twentieth century, photographic forgeries were widely used in political propaganda. Some noteworthy historical examples are:

- **Stalinist Russia (1930s–1950s):** Under Joseph Stalin's rule, photo manipulation became a powerful tool of political propaganda. Opponents who fell out of favor

1.3 Image Forgery

Fig. 1.2 Hippolyte Bayard's self-portrait (1840) (taken from https://en.wikipedia.org/wiki/Hippolyte_Bayard)

were not only purged from the Communist Party but were also literally erased from history by removing them from official photographs. These edits were made to rewrite historical records and create the illusion of Stalin's undisputed leadership. A famous example includes the erasure of Nikolai Yezhov, a high-ranking official, from a photograph after his execution (Fig. 1.3) [14].

- **Nazi Germany (1933–1945):** Adolf Hitler's regime also employed image manipulation to promote its ideology and leader cult. Photographs were retouched to enhance Hitler's presence at public events, and individuals deemed undesirable by the regime were removed or replaced. These doctored images were used in newspapers, books, and public exhibitions to reinforce Nazi propaganda [15].

These analog manipulations were performed in darkrooms using techniques like dodging, burning, airbrushing, and reprinting.

With the advent of digital cameras and scanners, images became digital i.e. sequences of bits, opening up entirely new manipulation possibilities. Software tools like Adobe Photoshop; professional image editing software developed by Adobe Inc., released in 1990 became a standard tool for digital image manipulation and is widely used in photography, graphic design, advertising, publishing, and art. Photoshop offers an extensive range of tools for image retouching, composition, color correction, and object removal, which makes it a creative image editing tool and therefore a potential medium for image forgery. Some famous examples of early digital image forgeries are:

Fig. 1.3 Stalin with and without Nikolai Yezhov, the chief of the NKVD, (1930s), Moscow-Volga Canal, Russian SFSR (taken from https://www.comradegallery.com/journal/fabrication-photographs-stalin-soviet-state?srsltid=AfmBOoo4aLgN4iYdN4c8fdADXrQhaHnBPgtrs3-72X6 oufIsUmsnkRJv)

- The Tourist Guy Hoax (2001) [16]: Shortly after the September 11 attacks at World trade center US, a photo began circulating via email, showing a man standing atop the World Trade Center, with a hijacked plane seemingly approaching in the background. This image, dubbed the "Tourist Guy", was later revealed to be a digitally manipulated image. The man in the photo was a tourist, and the image of the plane was superimposed using image editing software. The shadows and date inconsistencies quickly gave away the hoax. Nevertheless, it went viral, becoming one of the earliest examples of internet-based image forgery that spread widely and was even used for satirical memes.
- Time Magazine Cover Controversy (1994) [17]: In June 1994, Time magazine published a cover featuring a mugshot of O.J. Simpson, arrested for the murder of Nicole Brown Simpson and Ronald Goldman. The image was digitally manipulated to appear darker and more shadowed, giving it a more sinister tone. This sparked major backlash, with critics accusing Time of racial bias and unethical journalism. Newsweek ran the same mugshot unaltered, which made the manipulation even more apparent. The incident raised serious questions about the ethics of photo editing in news media.

These two examples highlight different types of image forgery:

- The Tourist Guy: A hoax with viral misinformation potential
- Time Magazine: Subtle tonal manipulation used to sway perception, especially in a racially charged context

Both cases show how image manipulation can distort reality, impact public opinion, and question the credibility of media outlets.

Digital tools allowed easy cropping, cloning, morphing, and retouching, drastically lowering the skill barrier for creating forged content.

1.3 Image Forgery

With the increase in use of social media platforms like Facebook, WhatsApp, X (formerly Twitter), and Instagram, manipulated images became widespread. These were often used for entertainment but also to spread misinformation, particularly during elections or social movements.

Reuters Lebanon War Photos (2006) (Fig. 1.4) controversy is another key example in the timeline of digital image forgery in journalism. In August 2006, during the Israel–Hezbollah conflict in Lebanon, Reuters was forced to withdraw two images taken by freelance photographer Adnan Hajj after it was discovered that they had been digitally manipulated using software. In one widely circulated photo, smoke rising over Beirut from an Israeli airstrike was cloned and darkened to exaggerate the scale of the destruction. In another image, flares from an Israeli jet were digitally enhanced and multiplied to make it appear more aggressive. The manipulations were exposed by bloggers and photography analysts who spotted repeating patterns in the smoke and artifacts from cloning tools. In the response, Reuters had to retract the altered images and terminating their relationship with the photographer. The case severely damaged trust in photojournalism during wartime reporting and sparked a debate on media ethics, photojournalism integrity, and the need for transparent editorial oversight in the age of digital photography [16, 18–20].

Professional photojournalism organizations began enforcing stricter ethical guidelines, and photo forensics began gaining traction.

Now a days, with the help of sophisticated image editing tools, many image manipulation forgery techniques have emerged, for example, copying and pasting

Fig. 1.4 Digitally altered photograph of the aftermath of an IDF attack on Beirut. (Reuters) https://en.wikipedia.org/wiki/File:Beirut-smoke.jpg

regions within the same image to conceal or duplicate objects (Copy-Move image forgery), combining objects from multiple images into a single composite (Image Splicing), Metadata Tampering.

The introduction of Generative Adversarial Networks (GANs) revolutionized image generation. This image generation added a new era in the image forgery. These networks can produce synthetic images that mimic the distribution of real ones [20]. Earlier, a digital image was required to create a fake image but now any real image is not required rather a photorealistic image (synthetic image/non-real image) or video can be generated and used for bad intention.

Image forgery created using readily available tools and techniques is also termed as "Shallowfae" or "Cheapfake", while forgery using complex and advanced AI to generate realistic but fake content is termed as deepfake.

Forged images whether manipulated or AI-generated images are used these days in Social media platforms and other areas. Modern image forgery detection techniques address many of these, but still this field remains a current topic of research.

1.4 Conclusion

From painted negatives to GAN-generated deepfakes, the history of image manipulation reflects a continuous interaction between innovation and deception. While early techniques were limited by skill and technology, modern tools enable even a novice user to create convincing forgeries. As manipulation becomes harder to detect, especially with AI advancements, the importance of robust forgery detection methods, ethical awareness, and policy interventions becomes more critical.

References

1. Kemp, M.: The Science of Art: Optical Themes in Western Art from Brunelleschi to Seurat. Yale University Press, United Kingdom (1990)
2. Gernsheim, H.: The Origins of Photography. Thames and Hudson, New York (1982)
3. Newhall, B.: The History of Photography: From 1839 to the Present. Museum of Modern Art, United Kingdom (1982)
4. Talbot, W.H.F.: The Pencil of Nature. Longman, Brown, Green & Longmans (1844) https://www.gutenberg.org/files/33447/33447-h/33447-h.html
5. Archer, F.S.: On the use of collodion in photography. The Chemist. **2** (1851)
6. Maddox, R.L.: An experiment with Gelatino-Bromide. Br. J. Photogr. **18**, 426–428 (1871)
7. Collins, D.: The Story of Kodak. Harry N Abrams Inc, New York (1990)
8. Reilly, J.B.: Care and Identification of 19th-Century Photographic Prints. Eastman Kodak Company (1986)
9. Land, E.H.: A new one-step photographic process. J. Opt. Soc. Am. **37**, 61–77 (1947)
10. Sasson, S.: History of the First Digital Camera. Kodak Historical Archives (2007)
11. Sarvas, R., Frohlich, D.: From Snapshots to Social Media—The Changing Picture of Domestic Photography. Springer, London (2011)

References

12. Sapir, M.: The impossible photograph: Hippolyte Bayard's self-portrait as a drowned man. MFS Mod. Fict. Stud. **40**(3), 619–629 (1994). https://doi.org/10.1353/mfs.1994.0007
13. Lester, P.M.: Photojournalism: An Ethical Approach, 1st edn. Routledge (1991). https://doi.org/10.4324/9781315682518
14. King, D.: The Commissar Vanishes: The Falsification of Photographs and Art in Stalin's Russia. Metropolitan Books (1997)
15. Strobl, G.: The Swastika and the Stage: German Theatre and Society, 1933–1945. Cambridge University Press (2008)
16. Perlmutter, D.: Photojournalism and Foreign Policy: Icons of Outrage in International Crises. (Praeger Series in Political Communication). Praeger Publishers (1998)
17. Zelizer, B.: When facts, truth, and reality are God-terms: on journalism's uneasy place in cultural studies. Commun. Crit./Cult. Stud. **1**(1), 100–119 (2004). https://doi.org/10.1080/1479142042000180953
18. Lewis, P.: Reuters drops photographer over doctored images. The Guardian [Online]. https://www.theguardian.com/media/2006/aug/07/reuters.pressandpublishing (2006, August 7)
19. Lind, D.M.: Photo manipulation and its effect on future credibility of documentary photojournalism. (1993). https://doi.org/10.26076/93d5-945e
20. Goodfellow, I., Pouget-Abadie, J., Mirza, M., Bing, X., Warde-Farley, D., Ozair, S., Courville, A., Bengio, Y.: Generative adversarial networks. Commun. ACM. **63**(11), 139–144 (2020). https://doi.org/10.1145/3422622

Chapter 2
Types of Digital Image Forgery

Abstract Digital image forgery is the process of manipulating or fabricating image content to mislead viewers or misrepresent reality. With the proliferation of image-editing tools and generative AI, image forgeries have become increasingly sophisticated and prevalent. This chapter categorizes digital image forgeries into two major classes: manipulated real images, which involve alterations of real images, and computer-generated images, which are synthetic images created from scratch using a software. This chapter provides an in-depth overview of the primary types of digital image forgeries.

Keywords Copy–move forgery · Image splicing · Inpainting · Computer generated image · Image forgery

2.1 Introduction

Digital images are integral to information sharing in journalism, social media, forensic science, and legal documentation. The authenticity of such images is critical as forged image may provide false and incorrect information. Forged images are often indistinguishable from authentic images by human eye.

With the ubiquitous use of digital images in journalism, social media, science, research and law, the potential for visual misinformation has increased dramatically. At the same time the number of tampered images is growing rapidly due to availability of low-cost or free image editing tools which are easy to use by any non-expert also.

Digital image forensics is a systematic attempt to assess the trustworthiness of a digital image. Two of the specific goals of digital image forensics are the detection of computer-generated images and the detection of processed photographic or forged images [1, 2].

Based on the type of images, image forgery can be divided into two categories:

(a) Manipulation in photographic digital images
(b) Computer generated images

2.2 Manipulation in Photographic Digital Images

Photographic images that have undergone some alteration are referred to as processed photographic digital images. In general, there are two types of image processing methods: legitimate (honest) processing and illegitimate (illegal) processing. Legitimate processing of images is done for cosmetic purposes which should be genuine, ethical, and lawfuland without altering semantic meaning or authenticity of the image [2]. For example, resizing an image or enhancing the visual quality of an image are legitimate image processing as these image processing operations are not changing semantic meaning of the image. On the other side, the semantic meaning of a photographic image is altered in illegitimate image processing and therefore authenticity becomes doubtful. Illegitimate processing originates from authentic sources but after processing the meaning or context is distorted. Copy–move and image splicing forgery are examples of illegitimate image processing.

2.2.1 Legitimate Image Processing

An image processing operation, if not done to mislead or falsify the information is legitimate (honest) image processing and is not image forgery. Following image processing operations does not constitute image forgery, if done without any wrong motives:

- **Image Enhancements for Visibility or Quality (Non-Deceptive)**
  Sometimes image processing operations are performed to improve the image visual quality or visibility enhancement without altering the content or context of original image. These operations, if done without any wrong motive and ethically as well as legally acceptable then are not considered as image forgery e.g. change in brightness, contrast, and sharpness, noise reduction or resolution enhancement, color correction. Such operations are very useful in applications like medical image processing, CCTV images analysis.
- **Image Compression and Format Conversion**
  To reduce the space taken by an image file, or to efficiently transmit image file over networks, or resizing image to address compatibility issues, a digital image need to be compressed or format change is done., for example, conversion to JPEG, PNG formats or resizing for web pages as per requirement. If such image processing operations are performed not to hide any information or any other wrong motive, then these image processing operations do not constitute image forgery.

- **Image Restoration or Reconstruction**
  Image processing operations aimed to restore an image by removing distortions or defects that occurred during image acquisition, transmission, or storage like Deblurring (removing motion or focus blur), scratch removal (fixing damaged or old photos), Dehazing (improving visibility in foggy or smoky scenes) or Inpainting (declared restoration: filling in missing or damaged parts from known surroundings) are not considered as image forgery if the motive is to recover the original appearance of the image and reconstruction, if any, is based on contextual or adjacent data, not based on the imagination of anyone. The aim of these image processing operations should not be misrepresentation.
- **Image Cropping or Rescaling (with context preservation)**
  If any image is cropped or rescaled for focus or some layout adjustment without removing and critical information from image or without any change in interpretation then it is not considered as image forgery.

Apart from these some other image edits like applying image processing filters, conversion into an artistic or stylized image, adding watermark for authenticity, automatic processing by digital camera like panoramastitching, color balancing, image inpainting are not considered as image forgery if contents are not falsified.

In conclusion, it can be said that the motive or intention is main criteria in determining whether an image processing operation can be considered as image forgery or not. Any image processing operation to improve clarity, preserve, or enhance the image without introducing false information or misleading the viewer is not an image forgery.

2.2.2 Illegitimate Image Processing or Image Forgery

Image processing operations performed involving intentional alterations in images which are aimed at deceiving, misinforming, or misrepresenting reality. These manipulations are performed without disclosure, constitute digital image forgery, carrying ethical, legal, and societal consequences and change semantic meaning of an image. Forgery in an image can be categorized in following types [3]:

- **Copy–Move Forgery**
  Copy–move forgery involves copying a region of an image and pasting it into another area of the same image. This type of forgery is performed to hide any object or replicate objects within same image. As lighting directions, illumination, texture, lighting, and noise copied and pasted regions are, from the same image, detection of copy–move image forgery becomes a challenging task. In copy–move forgery only one image is used. This type of manipulation alters the factual content of the image while maintaining visual consistency, making it hard to notice with the naked eye. When undisclosed, this image processing operation misleads viewers.
  Copy–move forgery also has various variants. When the copied region is simply translated at different locations within the image without any post-processing or

Fig. 2.1 Example of copy–move image forgery (CoMoFod database) (taken from https://www.vcl.fer.hr/comofod/examples.html)

transformation, this is commonly termed as plain or rigid copy–move forgery. The plain copy–move forgery may also include multiple regions of cloning. Plain copy–move forgery is the simplest form of copy–move forgery to be detected. The rotation and scaling are the two common image processing transforms that are applied to the copied regions before pasting (Fig. 2.1). Besides rotation and scaling transformations, various post-processing operations are also applied to forge the images. The post-processing operations include JPEG compression, noise addition, brightness change, color reduction, smoothing, contrast enhancement (Fig. 2.2). These post-processing operations can be applied glob-

2.2 Manipulation in Photographic Digital Images

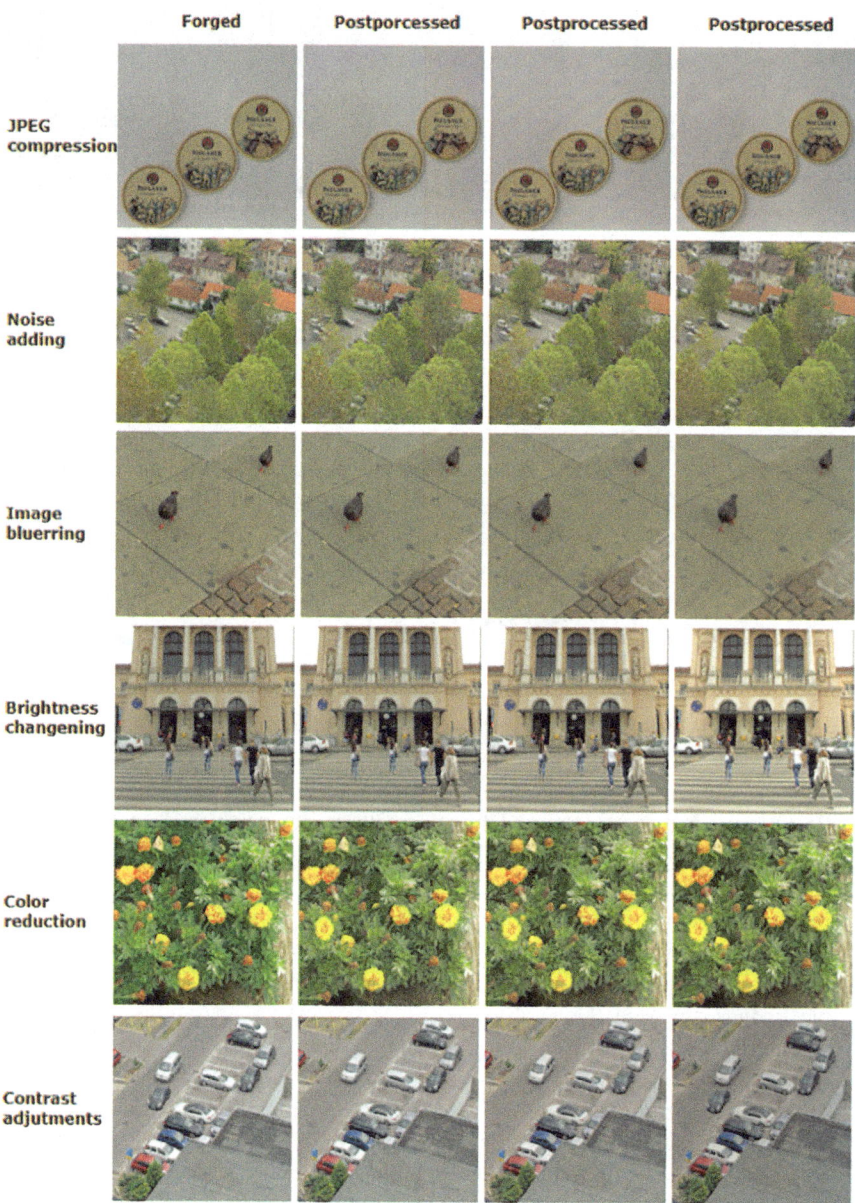

Fig. 2.2 Example of copy–move image forgery with post-processing (CoMoFod database) (taken from https://www.vcl.fer.hr/comofod/postprocessing.html)

Fig. 2.3 Example of image splicing forgery [3]

Fig. 2.4 Example of image inpainting forgery (**a**) Original image (**b**) Forged image [4]

ally on the whole forged image or locally only on the copied region. Post-processing operations and transformations are performed so that image forgery detection becomes difficult.

- **Image Splicing**

  In image splicing, regions from multiple images are combined into one composite image or spliced image (Fig. 2.3). This type of image forgery is used to show fabricated events or relationships in an image. In image splicing forgery; multiple images are used. Generally some post-processing operations are applied while creating image splicing forgery. For example, to hide the abrupt changes, image smoothing is performed near the boundary lines of the spliced images. As the size and resolution of source images may not be the same, therefore most of the time, a resampling (upsampling or downsampling) is performed to adjust the shape and sizes of different objects such that spliced image should appear more realistic.

- **Image enhancement/Image Retouching**

  Image enhancement operations as discussed in Sect. 2.2.1 (a) above are considered as digital image forgery, if done with wrong motive. This type of forgery is very commonly done in advertising industry to mislead the customers.

- **Image inpainting**

  Image inpainting is the process of removing an object from an image and filling in the region using surrounding texture (Fig. 2.4). Originally inpainting was used

for restoration of an image, but if used to manipulate content to remove some object from a scene, it is considered as image forgery [5, 6].
- **Meta data alteration**
If metadata (e.g. geographic location, time stamp, camera information) in an image is deleted or altered to conceal evidence or change narrative then it is termed as image forgery because it removes factual information of an image.

2.3 Computer Generated Images

The use of computer graphics has increased multifold in the last few years. Consequently, so many new doors have been opened for several technologies such as virtual reality, 3D gaming. Computer graphics-based tools are used in various paramount fields including the film industry for special effects, education, cartography, visualization of computer simulations, computer or mobile games, and medicine. As every technology has positive and negative aspects, similarly computer graphics also has a dark side. Computer-Generated Images (CGIs) are visuals created using computer graphics techniques rather than being acquired through a physical camera sensor. Unlike traditional photographic images, which capture real-world scenes using optical and electronic hardware, Computer generated images are synthesized from mathematical models, rendering engines, and increasingly, artificial intelligence (AI) algorithms. These images can represent either realistic scenes that closely mimic the appearance of real photographs or abstract graphics for visualizations, animations, and simulations [7] (Fig. 2.5). Computer generated images are also called synthetic images. These images often feature physically-based rendering (PBR) techniques, which simulate real-world lighting interactions to generate high-fidelity visuals.

Fig. 2.5 Examples of computer-generated images (Top row) and photographic images (Bottom row) taken from the DSTok dataset [5]

Computer-generated images with high photorealism may create social disorder when computer-generated images with false misleading information are created and shared over social media. The problem is becoming more and more dangerous with the advancement of computer graphics rendering tools. At present, images can be created with such photorealism that the human visual system can be easily fooled. Hence the proverb "seeing is believing" is no more true especially for digital images. Some of the recent inventions in the field of computer graphics include GAN (Generative Adversarial Network), Variational Autoencoders (VAEs), and Diffusion Models. These technologies can be used to create extremely realistic computer-generated images and videos in real-time. These images often lack certain physical traits found in real images such as Photo Response Non-Uniformity (PRNU), chromatic aberration, sensor noise, and lens distortions making traditional forensic methods based on these artifacts ineffective. Moreover, AI-generated images typically do not contain metadata, such as timestamp, GPS location, or camera model, which can be used for verifying image authenticity [4].

Another related challenge related to digital image forgery arises with hybrid images, where real photographs are partially manipulated using computer generated elements. For instance, inserting computer generated faces into group photos or blending rendered objects into real backgrounds makes detection even more complex.

In conclusion, while computer-generated images serve as powerful tools in design, simulation, and creativity, they also introduce new vulnerabilities in the digital content ecosystem. As computer generated images continue to evolve in realism and complexity, particularly through AI, the development of robust forensic techniques to differentiate them from authentic images is critical for preserving trust, particularly in legal, journalistic, and scientific applications.

2.4 Conclusion

Digital image forgery has evolved from simple cut-and-paste techniques to sophisticated synthetic content creation. Each type of forgery discussed in this chapter reflects different levels of technical complexity, intent, and societal impact. Understanding this taxonomy is essential for researchers, journalists, policymakers, and engineers who aim to ensure visual authenticity in the digital age.

References

1. Farid, H.: A survey of image forgery detection. IEEE Signal Process. Mag. **26**(2), 16–25 (2009)
2. Farid, H.: Digital image forensics. Sci. Am. **298**(6), 66–71 (2008)
3. Ansari, M.D., Ghrera, S.P., Tyagi, V.: Pixel-based image forgery detection: a review. IETE J. Educ. **55**(1), 40–46 (2014). https://doi.org/10.1080/09747338.2014.921415

References

4. Meena, K.B., Tyagi, V.: A Deep Learning Based Method to Discriminate Between Photorealistic Computer Generated Images and Photographic Images. Communications in Computer and Information Science, vol. 1244. Springer, Singapore (2020). https://doi.org/10.1007/978-981-15-6634-9_20
5. Trung, D.T., Beghdadi, A., Larabi, M.-C.: Blind Inpainting Forgery Detection. In: 2014 IEEE Global Conference on Signal and Information Processing (GlobalSIP), Atlanta, GA, USA, pp. 1019–1023. IEEE (2014). https://doi.org/10.1109/GlobalSIP.2014.7032275
6. Bertalmio, M., Sapiro, G., Caselles, V., Ballester, C.: Image inpainting. In: Proceedings of the 27th Annual Conference on Computer Graphics and Interactive Techniques (SIGGRAPH '00), pp. 417–424. ACM Press/Addison-Wesley Publishing (2000). https://doi.org/10.1145/344779.344972
7. Tokuda, E., Pedrini, H., Rocha, A.: Computer generated images vs. digital photographs: a synergetic feature and classifier combination approach. J. Vis. Commun. Image Represent. **24**, 1276–1292 (2013). https://doi.org/10.1016/j.jvcir.2013.08.009

Chapter 3
Digital Image Forgery Detection Techniques

Abstract This chapter provides an overview of digital image forgery detection techniques, categorized into active and passive approaches. Active methods, such as digital watermarking and digital signatures, require pre-embedded information, while passive methods analyze inconsistencies without prior data. Traditional passive techniques include pixel, format, camera, geometry, and environment-based methods. The chapter also highlights the rise of deep learning approaches using CNNs, Autoencoders, GANs, RNNs, Transformers, and hybrid models. It concludes by emphasizing the need for advanced, adaptable systems to address increasingly sophisticated forgeries.

Keywords Active image forgery detection · Passive image forgery detection · Deep learning algorithms · Copy–move forgery · Inpainting · Image splicing

3.1 Introduction

Digital images are common and popular form of information exchange in the digital age. However, ease of digital image manipulation has also posed serious threats to authenticity and trustworthiness. Forgery detection techniques aim to verify the authenticity and integrity of an image, especially in critical domains such as journalism, legal evidence, surveillance, and scientific imaging [1].

This chapter discusses image forgery detection techniques, which are used traditionally; primarily based on statistical analysis, signal processing, and handcrafted feature extraction and advanced deep learning based image forgery detection techniques.

3.2 Image Forgery Detection Techniques

Image forgery detection is the process of identifying whether a digital image has been tampered with and, if so, the tempered area in the image. It ensures the authenticity and integrity of visual data. Various forgery detection techniques have been suggested by the researchers ranging from traditional techniques to advanced deep learning based techniques (Fig. 3.1).

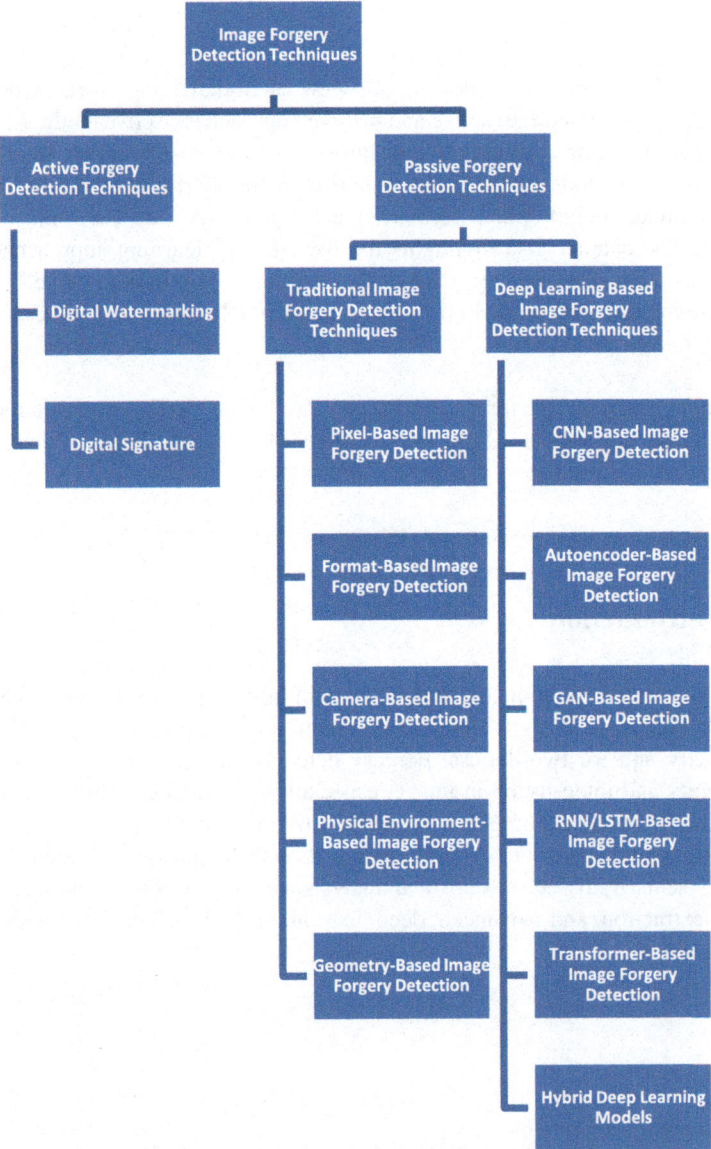

Fig. 3.1 Classification of digital image forgery detection techniques

Broadly, image forgery detection techniques are categorized into two main classes:

- Active image forgery detection techniques
- Passive image forgery detection techniques

These classes are described in sections below with various type of techniques in each class used for detection of image forgery.

3.3 Active Image Forgery Detection Techniques

In this type of forgery detection techniques, prior information is required to detect forgery. This information is embedded in the image itself at the time of image acquisition or creation, which is later used to detect tampering. In this type of techniques, if there is any suspicion of tampering, this pre-embedded information is extracted and used to detect image forgery.

Common techniques in active forgery detection category are:

- Digital watermarking [2–5]
- Digital signature [6]

Digital Watermarking In digital watermarking technique, a watermark (visible or invisible) is embedded into the image. Any manipulation in the image such as cropping, splicing, or retouching will disturb the watermark. Matching of this extracted watermark is done with original watermark to detect authenticity of the image. It is highly sensitive to changes, making it ideal for authenticity verification and tamper detection.

Watermarking is very useful in applications like Document Verification (e.g., scanned certificates, official images), Medical Imaging (Ensuring image integrity in diagnosis systems), Digital Forensics and copyright protection (for authenticity, not robustness). A review of various watermarking techniques used to detect image forgery can be found in literature in [7–9].

Digital Signature In digital signatures, the image is hashed, and the hash is encrypted with the sender's private key. At the receiver's end, the hash of the received image is compared with the decrypted hash. If the image is altered in any way, the hash or digital signature will not match, indicating possible tampering. It is used for source authentication and content integrity.

Digital signatures are very useful in applications like medical imaging to ensure diagnostic images are not tampered, authenticating scanned documents, IDs, etc.

Extensive research has been done in the area of digital signatures. References [10–14] can be read for further details.

3.4 Passive Image Forgery Detection Techniques

Passive image forgery detection (also known as blind forgery detection) techniques are used if there is no pre-embedded information available in the image [15]. These techniques exploit changes in natural properties like statistical, structural, or semantic inconsistencies of an image that arise due to tampering in the image.

Passive forgery detection techniques can be broadly classified as:

- Traditional Image Forgery Detection Techniques
- Traditional Computer generated Image Detection Techniques
- Deep learning based Image Forgery Detection Techniques

3.4.1 Traditional Image Forgery Detection Techniques

Traditional image forgery detection techniques do not rely on pre-embedded information like watermarks or signatures. Instead, these techniques analyze inner features of the image, such as inconsistencies in color, lighting, geometry, or compression artifacts, to detect manipulation. These techniques can be categorized as follows:

- Pixel-Based image forgery detection techniques
- Format-Based image forgery detection techniques
- Camera-Based image forgery detection techniques
- Physical environment based image forgery detection techniques
- Geometry based image forgery detection techniques

3.4.1.1 Pixel-Based Image Forgery Detection Techniques

Pixel-Based Image Forgery Detection techniques are under passive forgery detection category that inspects the pixel-level properties and patterns in a digital image to identify inconsistencies caused by tampering. Pixel-based image forgery detection techniques focus purely on statistical anomalies, duplication patterns, and compression artifacts directly within the image data itself introduced due to image forgery, without considering the image source or real-world physics. The analysis is performed based on pixel-level analysis. These techniques are passive detection techniques as the original image is not required to detect image forgery.

Pixel-Based Image Forgery Detection Techniques are used to detect following types of image forgery:

- **Copy–Move Forgery Detection (CMFD)**
  As discussed in Chap. 2, copy–move forgery involves copying a region from the image and pasting it elsewhere (to hide or duplicate content) within the source image itself. The copied and pasted image regions are usually visually similar

3.4 Passive Image Forgery Detection Techniques

with other parts of image, making forgery detection difficult, but may differ slightly due to scaling, rotation, or noise. This type of forgery can be detected by dividing the image into blocks or keypoints (like SIFT, SURF) [16–18, 40], and finding self-similar blocks in the image [19]. The block-based CMFD methods, in general, divide the image to be detected into small, regular, and overlapped blocks. Then features of each of these subblocks are extracted and then the results are obtained by matching and post-processing those features. These methods have high computational complexity. The keypoint-based CMFD methods usually extract features from an entire image, which is the main difference from block-based methods, and they effectively reduce computational complexity [20]. A number of copy–move forgery detection methods have been developed, some of these are given in [21].

- **Image Splicing Detection**
  Splicing image forgery involves combining parts from two or more images to create a forged composite image. As the image parts are taken from different images, pixel-based methods can detect boundary inconsistencies, color mismatches, or blending artifacts between spliced regions. Splicing detection techniques include edge detection, statistical feature analysis, and deep learning-based feature extraction. A survey of various image splicing detection is provided in [22, 23].

  In Copy–move image forgery detection, the main objective is to find similar regions within an image, while in Image splicing detection, the aim is to find distinguishing features between the spliced and authentic regions. The basic idea used in detection of image splicing is that the noise levels of source images used in splicing are different. Image splicing detection is more challenging in comparison to copy–move forgery detection. Following assumptions help in detecting image splicing: (1) Image is resampled during the process of image splicing and (2) Image retouching has been done to make sliced boundaries smooth to make splicing less visible.

- **Resampling Detection**
  Tampering often involves resampling or interpolation to estimate pixel values at various pixel positions. These interpolation processes introduce periodic correlations in the resampled image grid, leading to subtle but consistent artifacts in the frequency domain. Such periodic patterns, often imperceptible to the human eye, can be effectively detected using signal processing techniques such as the Fourier Transform or autocorrelation analysis. The underlying principle is that resampling introduces periodic structures that manifest as peaks in the frequency spectrum or in the autocorrelation map of image derivatives [24, 25].

- **JPEG Compression Artifact Analysis**
  JPEG is a widely used lossy compression standard that transforms image blocks (usually 8×8 pixels) into the frequency domain using the Discrete Cosine Transform (DCT), quantizes the coefficients, and then encodes them. This process introduces characteristic compression artifacts, especially at block boundaries, and leads to rounding effects in the DCT coefficient histograms.

When a tampered image is saved in JPEG format, especially when a region is manipulated and recompressed separately, the resulting image may contain inconsistent compression signatures across different regions. These inconsistencies form the basis of JPEG artifact analysis for forensic purposes.

When an image is tampered and compressed again, double JPEG compression often occurs. In such cases, analysis of JPEG blocking artifacts and quantization tables can reveal forgery. Inconsistencies in DCT coefficients help detect tampered regions [26].

Error level analysis that visualizes the compression error in different regions of the image can be used to detect image forgery. Forged areas in an image typically show higher error levels due to local editing and recompression, even when visually similar to surrounding pixels.

Performance of pixel based forgery detection techniques degrades on low-resolution, recompressed, or blurred images. These techniques may also produce false positives in textured regions of an image, and also vulnerable to anti-forensic techniques that smooth pixel anomalies.

3.4.1.2 Format-Based Image Forgery Detection Techniques

Format-based image forgery detection is a passive forgery detection approach that depends on analyzing the internal characteristics of image file formats, particularly compression artifacts, encoding parameters, and metadata structure, to uncover signs of image manipulation. Among various image formats, JPEG is the most studied due to its widespread use and unique compression mechanism.

Unlike active techniques that require prior embedded information (e.g., watermarks or signatures), format-based techniques exploit inconsistencies introduced during image acquisition, saving, editing, and re-compression to detect tampering. These techniques are especially powerful for detecting splicing, region pasting, and post-processing operations in JPEG images.

3.4.1.3 Camera-Based Image Forgery Detection Techniques

Digital cameras leave several intrinsic traces or artifacts in images during the acquisition process due to the design and functioning of their internal components. Camera-Based Image Forgery Detection is a passive image authentication approach that relies on identifying these unique traces or artifacts introduced by digital cameras during image acquisition. These traces include sensor noise patterns, lens distortions, color filter array (CFA) interpolation, and JPEG compression signatures, which are unique to a particular camera or camera model. If these intrinsic properties in an image are disrupted or manipulated, it can signal a possibility of image forgery [27].

These techniques are useful in detection forgery in real photographic images and cannot detect forgery in synthetic images which are not taken by camera.

3.4.1.4 Physical Environment-Based Image Forgery Detection Techniques

Physical Environment-based image forgery detection are based on real-world physical parameters captured within an image such as lighting, shadows, geometry, and camera parameters to detect tampering [28]. These types of techniques are under the category of passive forensic approach. These techniques are very useful in scientific and legal investigations to authentic evidences requiring physical realism like accident or crime scene images. These techniques work on principle of detection of inconsistency of illumination, shadow analysis based on light source and relative object positions, environment conditions (like fog, depth blur).

These techniques may not provide good results in case of low resolution or heavily compressed images, complex lighting or multiple light sources.

3.4.1.5 Geometry-Based Image Forgery Detection Techniques

Geometry-based image forgery detection techniques are based on geometric inconsistencies introduced during the forgery process. Unlike other techniques that analyze local features or sensor patterns, geometry-based forgery detection techniques analyze the underlying structure of the image such as perspective, vanishing points and 3D constraints to identify forged regions. These methods are particularly useful when image forgeries are skillfully crafted and cannot be detected at the pixel level. Objects in a scene follow consistent vanishing points and perspective rules. In an image manipulated objects often violate the geometric rules of perspective. This includes incorrect scale, angle, or proportions relative to the scene. By observing 3D geometry and perspective inconsistencies, image forgery can be detected [29, 30].

These techniques are very useful in splicing detection where parts of multiple images are combined into one, and in verification of scene consistency in applications like crime scene images.

3.4.2 *Traditional Computer Generated (Synthetic) Image Detection Techniques*

In last few years, a number of research papers have been published on techniques to identify computer generated images. In general, a common workflow of these techniques is:

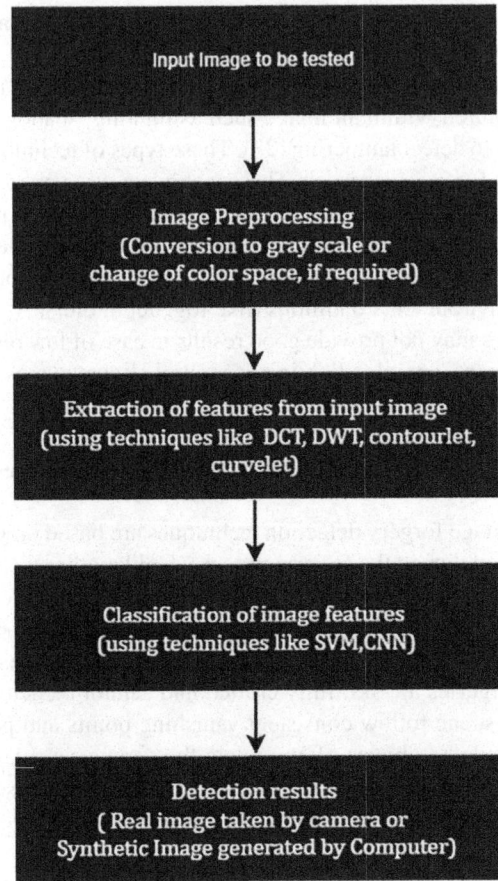

Image to be tested → Image Pre-processing (Conversion to gray scale or change of color space, if required) → extraction of features from input image (using techniques like, DCT, DWT, contourlet, curvelet) → Classification of image features (using techniques like SVM, CNN, FLD → Detection results (Real image taken by camera or Synthetic Image generated by computer).

Authors in [42] have categorized Computer generated image detection techniques as follows:

- Statistical distribution based techniques
- Visual-features based techniques
- Acquisition process based techniques
- Hybrid features based techniques
- Deep learning based techniques

3.4 Passive Image Forgery Detection Techniques 29

3.4.3 Deep Learning Based Image Forgery Detection Techniques

Image forgery detection techniques have evolved significantly with the advancements in deep learning techniques. Unlike traditional methods that rely on hand-crafted features, deep learning models can automatically learn rich representations from data, enabling more accurate and robust detection of complex forgeries. In this section, deep learning-based techniques for image forgery detection, including CNNs, Autoencoders, GANs, RNNs, Transformers, Hybrid Models, and Pixel-Level Localization methods are described.

3.4.3.1 Convolutional Neural Network (CNN)-Based Image Forgery Detection Techniques

CNNs are able to extract hierarchical features and therefore are very useful in many image analysis activities including image forgery detection. CNNs can be trained for binary classification (real image vs. Forged image) or multi-class classification (type of forgery). Patch-wise classification is often employed, where an image is divided into patches and each is evaluated for authenticity. Bayar and Stamm [31] introduced constrained convolutional layers designed to suppress image content and emphasize manipulation artifacts. Their approach demonstrated that carefully designed CNNs could distinguish subtle differences in authentic and manipulated image patches. In [32], a data-driven algorithm based on convolutional neural networks, which learns features characterizing each camera model directly from the acquired pictures is given.

Various CNN architectures used for image forgery detection are:

- VGGNet
- ResNet
- EfficientNet

3.4.3.2 Autoencoder-Based Image Forgery Detection Techniques

Autoencoders are unsupervised models that are particularly effective in anamoly detection applications. Autoencoders learn to compress and reconstruct images. During training, these models learn to reconstruct only the authentic patterns properly but when used with forged images, the reconstruction error is higher, especially in manipulated regions. In image forgery detection: forgery is localized by computing the pixel-wise difference between the input and the reconstructed image.

3.4.3.3 Generative Adversarial Network (GAN)-Based Image Forgery Detection Techniques

GANs are a class of deep learning models introduced by Ian Goodfellow et al. in 2014 [33], that are capable of generating realistic synthetic data, such as images, by learning the distribution of real data. GANs consist of two components:

- A generator that tries to produce fake images
- A discriminator that tries to distinguish real images from fake images

For image detection purposes, the discriminator can be trained to identify GAN-generated synthetic images or tampered images. Alternatively, adversarial training can be used to improve the robustness of detection models.

Zhou et al. [34] trained discriminators specifically for detecting DeepFakes. They also proposed ForensicTransfer, a domain adaptation framework using GANs for cross-dataset generalization.

GAN-based methods also serve as a strong baseline for generating challenging forged samples during training.

3.4.3.4 Recurrent Neural Networks (RNNs) and LSTM-Based Image Forgery Detection Techniques

Though primarily used for sequential data, RNNs and LSTMs are used in image forgery detection techniques in the following ways:

- Inter-patch dependency learning: Modeling spatial sequence of patches extracted from images
- Frame-level analysis in video forgery detection: Detecting frame inconsistencies using temporal modeling

Combined with CNNs, RNNs enhance context modeling, improving performance in cases where forgery involves sequential or structured changes [35].

3.4.3.5 Transformer-Based Image Forgery Detection Techniques

Transformers, originally introduced for natural language processing (NLP) have recently achieved remarkable success in computer vision, including image forgery detection. Their ability to model long-range dependencies makes them especially powerful in detecting subtle inconsistencies caused by forgery. Self-attention mechanism allows the model to focus on both local and global image contexts.

Commonly used transformers for forgery detection are:

- Vision Transformer (ViT): These are Patch-based techniques and divide an image into fixed-size patches and treats them like words in NLP. These can be used to

3.4 Passive Image Forgery Detection Techniques

detect image manipulations such as copy–move, splicing, or GAN-based forgeries [36]

- Swin Transformer: These are hierarchical architecture and use shifted windows for efficient computation. These are effective in both global and local forgery localization [37]

These models typically require large-scale datasets for training but yield superior results in challenging cases such as GAN-generated images. Other models used in image forgery detection are DETR (DEtection TRansformer) [39] and XCiT (Cross-Covariance Image Transformer) [38].

In [41], author has provided a generalized framework of digital image forgery detection techniques (Fig. 3.2).

3.4.3.6 Hybrid Deep Learning Image Forgery Detection Models

Hybrid deep learning models combine the strengths of multiple architectures discussed above, such as CNNs, RNNs, Autoencoders, GANs, and Transformers, to create more powerful and adaptable systems for image forgery detection and localization. These models aim to overcome the limitations of individual methods by integrating local feature extraction, temporal/spatial dependencies, and semantic context. Combining different architectures utilize the strengths of each and are very useful in detecting Multi-type forgery scenarios (splicing, copy–move) and GAN-generated, deepfake images.

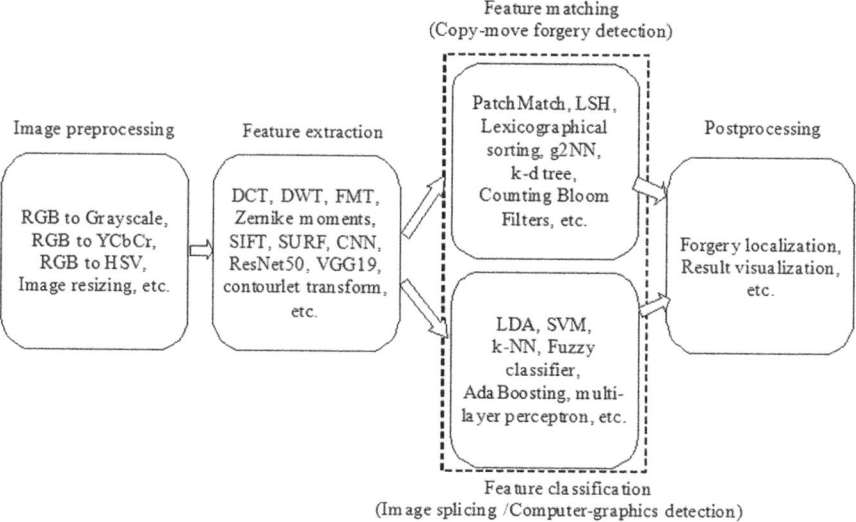

Fig. 3.2 A generalized framework of digital image forgery detection techniques

For example:

- **CNN + Attention**: The CNN extracts features, while attention layers highlight regions of interest
- **CNN + Transformer**: Provides both spatial feature extraction and global context modeling
- **CNN + LSTM**: Extracts spatial features and models their sequential dependencies

Hybrid models have shown improvements in accuracy and localization precision over single-architecture models.

3.4.3.7 Pixel-Level Image Forgery Localization Networks

Pixel-level forgery localization aims not just to detect if an image is forged, but exactly where the forgery has occurred at the pixel level. Unlike image-level classifiers that give a binary decision (real/fake), these networks output binary or probabilistic masks that highlight tampered regions. Localization requires detecting which pixels have been manipulated [14]. This task is typically addressed using:

- U-Net: An encoder-decoder architecture with skip connections
- Mask R-CNN: Detects object-like forged regions and creates binary masks
- Fully Convolutional Networks (FCNs): Provide dense pixel-wise classification

These networks are trained using ground-truth masks of forged regions, making them highly effective in accurately marking forged areas.

3.5 Conclusion

Image forgery detection is very essential in preserving the authenticity and trustworthiness of digital images across various fields. This chapter outlined both active and passive detection methods. While active methods provide robust authentication when prior information is available, passive methods offer flexibility by analyzing image content without pre-embedded data. The integration of deep learning models, such as CNNs, GANs, Transformers, and hybrid architectures, has significantly improved detection accuracy and localization capabilities. A bibliography of various forgery detection techniques is provided at the end of the book. As forgery techniques continue to evolve, future efforts must focus on developing more robust and generalizable models capable of adapting to new and complex manipulation methods.

References

1. Farid, H.: Image forgery detection. IEEE Signal Process. Mag. **26**(2), 16–25 (2009). https://doi.org/10.1109/MSP.2008.931079

References

2. Cox, J., Miller, M.L., Bloom, J.A.: Digital Watermarking. Morgan Kaufmann, San Francisco, CA (2002)
3. Wolfgang, R.B., Podilchuk, C.I., Delp, E.J.: Perceptual watermarks for digital images and video. Proc. IEEE. **87**(7), 1108–1126 (1999). https://doi.org/10.1109/5.771066
4. Barni, M., Bartolini, F., Piva, A.: Improved wavelet-based watermarking through pixel-wise masking. IEEE Trans. Image Process. **10**(5), 783–791 (2001). https://doi.org/10.1109/83.918566
5. Hsu, C.T., Wu, J.L.: Hidden digital watermarks in images. IEEE Trans. Image Process. **8**(1), 58–68 (1999). https://doi.org/10.1109/83.736696
6. Schneider, M., Chang, S.-F.: A robust content-based digital signature for image authentication. In: Proceedings of 3rd IEEE International Conference on Image Processing, pp. 227–230. IEEE (1996). https://doi.org/10.1109/ICIP.1996.560499
7. Alginahi, M., Hossain, M.S., Hossain, M.A.: A comprehensive survey on robust image watermarking: fundamentals, recent advances, and future directions. Neurocomputing. **492**, 312–337 (2022). https://doi.org/10.1016/j.neucom.2022.03.049
8. Zhang, R., Ma, Y., Liu, Y.: Deep learning-based watermarking techniques: a survey of state-of-the-art and future challenges. Circuits Syst. Signal Process. **43**, 1109–1133 (2024). https://doi.org/10.1007/s00034-024-02651-z
9. Zhang, Y., Wu, X., Zhao, J.: A survey on digital watermarking technology for AI-generated images. Mathematics. **13**(1), 1–18 (2025). https://doi.org/10.3390/math13010123
10. Zhang, H.-B., Yang, C., Quan, X.-M.: Image authentication based on digital signature and semi-fragile watermarking. J. Comput. Sci. Technol. **19**(6), 689–696 (2004)
11. Saad, S.M.: Design of a robust and secure digital signature scheme for image authentication over wireless channels. IET Inf. Secur. **3**(1), 45–54 (2009)
12. Li, Y., Wei, C.-H.: Digital image authentication: a review. Int. J. Digit. Libr. Syst. **2**(2), 1–24 (2011). https://doi.org/10.4018/jdls.2011040104
13. David, D., Divya, B.: Image authentication techniques and advances survey. COMPUSOFT Int. J. Adv. Comput. Technol. **4**(04), 1597–1601 (2024)
14. Jing, R.: A digital signature and watermarking based authentication system for JPEG2000 images. M.S. thesis, New Jersey Inst. Technol., Newark, NJ, USA (2005, January)
15. Ansari, M.D., Ghrera, S.P., Tyagi, V.: Pixel-based image forgery detection: a review. IETE J. Educ. **55**(1), 40–46 (2014). https://doi.org/10.1080/09747338.2014.921415
16. Mahdian, B., Saic, S.: Detection of copy–move forgery using a method based on blur moment invariants. Forensic Sci. Int. **171**, 180–189 (2007)
17. Huang, H., Guo, W., Zhang, Y.: Detection of copy-move forgery in digital images using SIFT algorithm. In: 2008 IEEE Pacific-Asia Workshop on Computational Intelligence and Industrial Application, Wuhan, China, pp. 272–276. IEEE (2008). https://doi.org/10.1109/PACIIA.2008.240
18. Bo, X., Junwen, W., Guangjie, L., Yuewei, D.: Image copy-move forgery detection based on SURF. In: 2010 International Conference on Multimedia Information Networking and Security, Nanjing, China, pp. 889–892. IEEE (2010). https://doi.org/10.1109/MINES.2010.189
19. Meena, K.B., Tyagi, V.: A copy-move image forgery detection technique based on tetrolet transform. J. Inf. Secur. Appl. **52**, 102481 (2020). https://doi.org/10.1016/j.jisa.2020.102481
20. Bensaad, A., Loukhaoukha, K., Sadoudi, S.: Keypoint-based copy-move forgery detection in digital images: a survey. In: 2022 7th International Conference on Image and Signal Processing and their Applications (ISPA), Mostaganem, Algeria, pp. 1–6. IEEE (2022). https://doi.org/10.1109/ISPA54004.2022.9786359
21. Anushree, R., Vinay Kumar, S.B., Sachin, B.M.: A survey on copy move forgery detection (CMFD) technique. In: 2023 International Conference on Intelligent and Innovative Technologies in Computing, Electrical and Electronics (IITCEE), Bengaluru, India, pp. 439–443. IEEE (2023). https://doi.org/10.1109/IITCEE57236.2023.10090884
22. Meena, K.B., Tyagi, V.: Image splicing forgery detection techniques: a review, advances in computing and data sciences. In: ICACDS 2021. Communications in Computer and Information Science, vol. 1441. Springer, Cham (2021). https://doi.org/10.1007/978-3-030-88244-0_35
23. Kumari, R., Garg, H.: Image splicing forgery detection: a review. Multimed. Tools Appl. **84**, 4163–4201 (2025). https://doi.org/10.1007/s11042-024-18801-z

24. Prasad, S., Ramakrishnan, K.R.: On resampling detection and its application to detect image tampering. In: 2006 IEEE International Conference on Multimedia and Expo, Toronto, ON, Canada, pp. 1325–1328. IEEE (2006). https://doi.org/10.1109/ICME.2006.262783
25. Popescu, A.C., Farid, H.: Exposing digital forgeries by detecting traces of resampling. IEEE Trans. Signal Process. **53**(2), 758–767 (2005)
26. Luo, W., Huang, J., Qiu, G.: JPEG error analysis and its applications to digital image forensics. IEEE Trans. Inf. Forensics Secur. **5**(3), 480–491 (2010)
27. Ho, A.T.S., Li, S.: Camera-based image forgery detection. In: Handbook of Digital Forensics of Multimedia Data and Devices, pp. 522–571. Wiley-IEEE (2015). https://doi.org/10.1002/9781118705773.ch14
28. Peterson, L.M., Kersten, D.J., Mannion, D.J.: Estimating lighting direction in scenes with multiple objects. Atten. Percept. Psychophys. **86**, 186–212 (2024). https://doi.org/10.3758/s13414-023-02718-0
29. Kee, E., O'Brien, J.F., Farid, H.: Exposing photo manipulation from shading and shadows. ACM Trans. Graph. **32**(3), 1–11 (2013)
30. Gallagher, C.: Detection of linear perspective manipulation in images. IEEE Trans. Inf. Forensics Secur. **5**(4), 833–841 (2010)
31. Bayar, B., Stamm, M.C.: A deep learning approach to universal image manipulation detection using a new convolutional layer. In: Proceedings of the 4th ACM Workshop on Information Hiding and Multimedia Security (IH&MMSec '16), pp. 5–10. Association for Computing Machinery (2016). https://doi.org/10.1145/2909827.2930786
32. Bondi, L., Baroffio, L., Güera, D., Bestagini, P., Delp, E.J., Tubaro, S.: First steps toward camera model identification with convolutional neural networks. IEEE Signal Process. Lett. **24**(3), 259–263 (2017). https://doi.org/10.1109/LSP.2016.2641006
33. Goodfellow, I.J., et al.: Generative adversarial nets. In: Advances in Neural Information Processing Systems, pp. 2672–2680. Curran Associates (2014)
34. Zhou, P., Han, X., Morariu, V.I., Davis, L.S.: Learning rich features for image manipulation detection. In: Proc. IEEE Conf. Comput. Vis. Pattern Recognit. (CVPR), pp. 1053–1061. IEEE (2018). https://doi.org/10.1109/CVPR.2018.00116
35. Bappy, J.H., Simons, C., Nataraj, L., Manjunath, B.S., Roy-Chowdhury, A.K.: Hybrid LSTM and encoder-decoder architecture for detection of image forgeries. IEEE Trans. Image Process. **28**(7), 3286–3300 (2019). https://doi.org/10.1109/TIP.2019.2895466
36. Huo, Y., Jin, K., Cai, J., Xiong, H., Pang, J.: Vision transformer (ViT)-based applications in image classification. In: 2023 IEEE 9th Intl Conference on Big Data Security on Cloud (Big Data Security), IEEE Intl Conference on High Performance and Smart Computing, (HPSC) and IEEE Intl Conference on Intelligent Data and Security (IDS), NY, USA, pp. 135–140. IEEE (2023). https://doi.org/10.1109/BigDataSecurity-HPSC-IDS58521.2023.00033
37. Liu, Z., et al.: Swin transformer: hierarchical vision transformer using shifted windows. In: 2021 IEEE/CVF International Conference on Computer Vision (ICCV), Montreal, QC, Canada, pp. 9992–10002. IEEE (2021). https://doi.org/10.1109/ICCV48922.2021.00986
38. Touvron, H., Cord, M., Douze, M., Massa, F., Sablayrolles, A., Jégou, H.: XCiT: cross-covariance image transformers. Proc. Adv. Neural Inf. Process. Syst. **34**, 20014–20027 (2021)
39. Zhu, X., Su, W., Lu, L., et al.: Deformable DETR: deformable transformers for end-to-end object detection. arXiv. (2020). https://doi.org/10.48550/arXiv.2010.04159
40. Lu, S., Hu, X., Wang, C., et al.: Copy-move image forgery detection based on evolving circular domains coverage. Multimed. Tools Appl. **81**, 37847–37872 (2022). https://doi.org/10.1007/s11042-022-12755-w
41. Meena, K.B.: Efficient passive forgery detection in digital images. Thesis. http://hdl.handle.net/10603/338230 (2021)
42. Meena, K.B., Tyagi, V.: Methods to distinguish photorealistic computer generated images from photographic images: a review. In: Advances in Computing and Data Sciences. ICACDS 2019. Communications in Computer and Information Science, vol. 1045. Springer, Singapore (2019). https://doi.org/10.1007/978-981-13-9939-8_7

Chapter 4
Datasets and Evaluation

Abstract In this chapter, a comprehensive overview of benchmark datasets and evaluation metrics used for assessing image forgery detection techniques is provided. It categorizes various publicly available datasets based on the type of forgery—such as copy-move, splicing, and synthetic image generation—and provides detailed descriptions of each dataset. Key datasets like CoMoFoD, CASIA TIDE, MICC, CISDE, ForgeryNet, and DeepGuardDB are discussed to highlight their significance in training and validating forgery detection algorithms. In addition, the chapter outlines evaluation metrics tailored to pixel-level, image-level, and region-level detection strategies.

Keywords Image forgery detection · Benchmark datasets · Evaluation metrics · Deepfake detection · Performance evaluation

4.1 Introduction

A digital image forgery detection system is trained and tested by performing experiments on benchmark databases. It helps in comparing performance of various forgery detection techniques. Some commonly used databases along with their details, used by researchers to train and test their proposed techniques in the area of Image forgery detection are provided in this chapter. Evaluation metrics are quantitative measures used to assess the performance of a model or detection algorithm. In the context of image forgery detection, they help determine how effectively a technique can identify forged vs. authentic images. This chapter also describes various evaluation metrics used to evaluate the performance of a technique used to detect image forgery.

4.2 Datasets Used for Training and Testing of Image Forgery Detection Techniques

As discussed in previous chapters, various types of forgeries can be done in digital images. Technique developed for these types require a database to training and testing its effectiveness. The commonly used databases are categorized according to forgery type in sections below:

4.2.1 Copy–Move Forgery Detection Related Datasets

(a) **CoMoFoD (Copy–Move Forgery Detection) Dataset [1]:**

The CoMoFoD dataset was created by researchers from the Faculty of Electrical Engineering and Computing at the University of Zagreb, Croatia. It comprises 260 sets of forged images. Each set includes a forged image, the corresponding original image, and two associated masks. The images are categorized into five types based on the manipulation technique used: translation, rotation, scaling, combination, and distortion. Additionally, post-processing operations such as JPEG compression, blurring, noise addition, and color reduction have been applied to both forged and original images. The dataset contains images in two different sizes:

- Small Image Category: This category includes 200 image sets, each with a resolution of 512 × 512 pixels, comprising 40 sets for each type of transformation. Including post-processed versions, the total number of images in this category is 10,400.
- Large Image Category: This category consists of 60 image sets with a resolution of 3000 × 2000 pixels, containing 10–20 sets for each type of transformation. The total number of images, including post-processed versions, is 3120.

CoMoFoD dataset can be downloaded from https://www.vcl.fer.hr/comofod/download.html

(b) **CMFD (Copy–Move Forgery Detection) Dataset [2]:**

CMFD dataset was developed by researchers at the Pattern Recognition Lab of Friedrich-Alexander-Universität Erlangen-Nürnberg (FAU), Germany. It comprises 48 original source images, each manually annotated with semantically meaningful regions referred to as snippets, that are intended for copy–move operations. These snippets are categorized into three types based on their content: smooth (e.g., sky), rough (e.g., rocks), and structured (typically man-made structures like buildings). A total of 87 such snippets are included.

To simulate realistic forgery conditions, various distortions such as JPEG compression artifacts, added noise, scaling, and rotation are introduced. A pixel-wise ground truth mask is also generated for each tampered image. The average resolution of images in this dataset is approximately 3000 × 2300 pixels, with tampered regions typically occupying around 10% of the image area.

4.2 Datasets Used for Training and Testing of Image Forgery Detection Techniques

An accompanying software tool allows users to dynamically create forged images by combining original images with tampered regions. This tool also automatically generates the corresponding ground truth for each manipulated image. It supports a range of modifications to the snippets, including Gaussian noise addition and affine transformations.

CMFDA dataset can be downloaded from https://www5.cs.fau.de/research/data/image-manipulation/

(c) **CMH (Copy–Move Hard) Dataset [3]:**

This dataset was created by researchers at the RECOD Lab, Institute of Computing, University of Campinas, located at Av. Albert Einstein, 1251, Cidade Universitária "Zeferino Vaz", Campinas, SP 13083-852, Brazil. Named CMH (Copy–Move Hard), the dataset is designed to present challenging scenarios for detecting copy–move forgeries.

It consists of 108 realistic cloned images in PNG format, with resolutions ranging from 845 × 634 to 1296 × 972 pixels. The dataset is divided into four categories:

- CMHP1: 23 images in which the cloned region was simply copied and moved (basic forgery);
- CMHP2: 25 images where the duplicated area was rotated, with angles ranging from −90° to 180°;
- CMHP3: 25 images where the cloned region was resized, with scaling factors between 80% and 154%;
- CMHP4: 35 images involving both rotation and scaling of the cloned region.

Each image is accompanied by a binary ground truth map highlighting the original and duplicated regions in white. To simulate compression effects, the dataset is also available in JPEG format, compressed at quality levels of 70%, 80%, and 90%. The final dataset includes a total of 216 images—half in the original uncompressed format and half in compressed JPEG format.

CMH Dataset can be downloaded from: https://doi.org/10.6084/m9.figshare.978736

(d) **MICC (F220, F2000, F8multi, F600) Datasets [4]:**

These datasets are developed by researchers at Media Integration and Communication Center (MICC) at the University of Florence, Italy and hence the name MICC. These four datasets contains images as follows:

- MICC-F220: contains 220 images total, out of these 110 images are tampered and 110 images are originals.
- MICC-F2000: contains 2000 images total, out of these 700 images are tampered and 1300 images are originals.
- MICC-F8multi: contains total eight tampered images having realistic multiple cloning.
- MICC-F600: contains 440 original images, 160 tampered images and 160 ground truth images.

MICC datasets can be downloaded from:

- http://lci.micc.unifi.it/labd/cmfd/MICC-F220.zip
- http://lci.micc.unifi.it/labd/cmfd/MICC-F2000.zip
- http://lci.micc.unifi.it/labd/cmfd/MICC-F8multi.zip
- http://lci.micc.unifi.it/labd/cmfd/MICC_F600.zip

(e) **CASIA TIDE (Chinese Academy of Sciences Institute of Automation—Tampered Image Detection Evaluation) V2.0 Dataset [5]:**

This dataset is developed by researchers at Chinese Academy of Sciences Institute of Automation and is an enhanced version of the original CASIA TIDE v1.0. This dataset can be used to evaluate: Copy–move forgery detection as well as Splicing detection algorithms.

This dataset can be downloaded from https://github.com/namtpham/casia2groundtruth

(f) **Forgery Image Dataset [6]:**

This dataset is developed by researchers at London South Bank University. This dataset consists of 1000 original and 3000 forgery images which are generated from the original images retrieved from publicly available repositories. Cut-paste, copy–move, and erase-filling types of forgery has been done in forged images. Both pre-processing operations (resizing, sharpening, color enhancement, blurring, regulating exposure) and post-processing operations (sampling, rotation, masking) have been applied for the generation of the forged images.

This dataset can be downloaded from https://ieee-dataport.org/documents/forgery-image-dataset

(g) **GRIP Dataset [7]:**

This dataset is developed by researchers at GRIP, an Image Processing Research Group of the University Federico II of Naples. Research interests of the group cover methodologies and applications of image processing and computer vision, with special emphasis on deep learning-based methods. Database composed by 80 images, having realistic copy–move forgeries, and size 768 × 1024 pixels. The forgeries are of arbitrary shapes, aimed at obtaining visually satisfactory results, with size going from about 4000 pixel (<1% of the image) to about 50,000 pixels.

This dataset can be downloaded from https://www.grip.unina.it/download/prog/CMFD/

4.2.2 Image Splicing Forgery Detection Related Datasets

(a) **CISDE (Columbia Image Splicing Detection Evaluation) Dataset [8]:**

This dataset is developed by researchers at Columbia University, New York. This dataset has 933 authentic images and 912 spliced image blocks of size 128 × 128 pixels. The image blocks are extracted from images in CalPhotos image set.

4.2 Datasets Used for Training and Testing of Image Forgery Detection Techniques

The complete details of the dataset are available at:
https://www.ee.columbia.edu/ln/dvmm/downloads/AuthSplicedDataSet/detailed.htm

(b) **MFC Datasets [9]:**
These datasets are provided by National Institute of Standards and Technology (NIST) Gaithersburg, Maryland. These datasets are provided as a benchmark for digital Media Forensics Challenge (MFC) evaluations. The data comprises over 176,000 high provenance (HP) images, more than 100,000 manipulated images and 35 million internet images. This dataset also provides 11,000 HP videos, 4000 manipulated videos; and 300,000 video clips.

(c) **CASIA TIDE (Chinese Academy of Sciences Institute of Automation—Tampered Image Detection Evaluation) V2.0 Dataset [5]:**
Details are provided in Sect. 4.2.1 (e) above.

4.2.3 Datasets to Evaluate an Algorithm for Photographic (Real) Image and Computer Generated (Synthetic) Image

(a) **DSTok (Deepfake and Synthetic Token) Dataset [10]:**
This dataset is provided by researchers at Institute of Computing, University of Campinas (Unicamp), Brazil. The DSTok dataset contains a variety of computer generated images. This dataset is having 4850 computer-generated and 4850 photographic images in JPEG format. The images in the dataset are of varying physical sizes in the range of 12 KB to 1.8 MB. The dataset contains computer-generated images of various resolutions ranging from 557 × 591 pixels to 3507 × 2737 pixels. The smallest photographic image is comprised of 800 × 600 pixels and the largest photographic image is of size 1800 × 1200 pixels. The computer-generated images available in this dataset are highly photorealistic. For very large images, the authors have cropped the central part.

(b) **Columbia Photographic Images and Photorealistic Computer Graphics Dataset [11]:**
This dataset is developed by researchers at Columbia University, New York. The dataset is composed of following four component image sets:

1. Photorealistic Computer Graphics Set
2. Personal Photographic Image Set
3. Google Image Set
4. Recaptured Computer Graphics Set

This dataset can be downloaded from:
https://www.ee.columbia.edu/ln/dvmm/downloads/PIM_PRCG_dataset/

(c) **DeepGuardDB: Real and Text-to-Image Synthetic Images Dataset [12]:**
The dataset consists of a total of 13,000 images, out of these images 6500 are AI-generated (fake) images and the remaining 6500 are authentic (real) images.

In addition to the images, a comprehensive JSON file that maps each pair of images (fake and real) to its corresponding prompts and associated hyper parameters is provided. This JSON file allows users to trace the generation process and understand the input parameters used for both the real and synthetic images, making the dataset more transparent and easier to analyse.

This dataset can be downloaded from:

https://ieee-dataport.org/documents/deepguarddb-real-and-text-image-synthetic-images-dataset#files

(d) **ForgeryNet: A Versatile Benchmark for Comprehensive Forgery Analysis [13]:**

ForgeryNet dataset contains 2.9 million images and 221,247 videos, seven image-level approaches and eight video-level approaches manipulations, perturbations (36 in-dependent and more mixed perturbations) and annotations (6.3 million classification labels, 2.9 million manipulated area annotations and 221,247 temporal forgery segment labels) from all over the world.

ForgeryNet dataset, an extremely large face forgery dataset with unified annotations in image- and video-level data across four tasks:

1. Image Forgery Classification, including two-way (real/fake), three-way (real/fake with identity-replaced forgery approaches/fake with identity-remained forgery approaches), and n-way (real and 15 respective forgery approaches) classification.
2. Spatial Forgery Localization, which segments the manipulated area of fake images compared to their corresponding real images.
3. Video Forgery Classification, which redefines the video-level forgery classification with manipulated frames in random positions. This task is important because attackers in real world are free to manipulate any target frame. and
4. Temporal Forgery Localization, to localize the temporal segments which are manipulated. ForgeryNet is by far the largest publicly available deep face forgery dataset in terms of data-scale

This dataset can be downloaded from: https://yinanhe.github.io/projects/forgerynet.html

(e) **DFFD: Diverse Fake Face Dataset [14]:**

This dataset is developed by a group of researchers at Computer vision lav, Michigan State University, The DFFD dataset is comprised of multiple publically available datasets and images that are synthesized/manipulated using publically available methods. By incorporating multiple sources for real images, this database contains images of varying resolution and image quality for both real and synthetic/manipulated images.

This dataset can be downloaded from: http://cvlab.cse.msu.edu/dffd-dataset.html

(f) **IMD 2020 [15]:**

This dataset is provided by researchers at The Czech Academy of Sciences, Institute of Information Theory and Automation, Prague, Czechia. In this work researchers have provided two large-scale and diverse datasets with a high variety of artifacts. A dataset of 35,000 real images captured by 2322 camera models are taken and the same number of digitally manipulated images by using a large variety of core image manipulation methods as well we advanced ones such as GAN or Inpainting resulting in a dataset of 70,000 images.

In addition to this dataset, a dataset of 2000 "real-life" (uncontrolled) manipulated images is also provided.

Both datasets can be downloaded from: http://staff.utia.cas.cz/novozada/db

4.3 Evaluation of Image Forgery Detection Techniques

To evaluate the performance of a forgery detection method, the selection of reliable metrics is an essential requirement. In various research papers, different types of evaluation metrics have been used by the authors. Digital image forgery detection algorithms can be divided into three categories:

(a) Pixel level image forgery detection
(b) Image level image forgery detection
(c) Region level image forgery detection

Each of these approaches serves a different purpose in the forgery analysis pipeline and employs different algorithms and evaluation metrics.

Pixel level image forgery detection algorithms produce a binary or probabilistic mask that indicates manipulated pixels within the image. These techniques are particularly effective for detecting fine-grained forgeries such as copy–move, splicing, or inpainting.

Image level image forgery detection algorithms are used in binary classification tasks. These methods do not localize the forgery but instead output a binary decision regarding the image's originality.

Region-level image forgery detection algorithms are in between image-level and pixel-level algorithms. These techniques aim to identify tampered regions in an image as distinct objects or connected components, without necessarily labeling each pixel.

Commonly used evaluation metrics for each level of algorithms are:

4.3.1 Pixel-Level Metrics

Pixel-level metrics evaluate how accurately a forgery detection algorithm can identify forged vs. authentic pixels in an image. These metrics compare the predicted pixel-wise binary mask (output by the algorithm) with the ground-truth mask (true locations of forged pixels). These metrics assess the accuracy of detecting forged pixels within an image. These metrics are:

(a) True Positives (TP): Number of correctly detected forged pixels.
(b) False Positives (FP): Number of genuine pixels incorrectly marked as forged.
(c) False Negatives (FN): Number of forged pixels not detected by the algorithm.
(d) True Negatives (TN): Number of genuine pixels correctly marked as genuine.

From these values, the following metrics are derived:

(a) Precision (positive predictive value): Measures how many detected forged pixels are actually forged.
(b) Recall (Sensitivity/True Positive Rate): Measures how many forged pixels were correctly detected.
(c) F1-Score: Harmonic mean of precision and recall, providing a balance between the two.
(d) Specificity (True Negative Rate): Measures how well genuine pixels are correctly identified.
(e) Accuracy: Measures the overall correctness of detection.
(f) Intersection over Union (IoU): Also known as the Jaccard Index; it measures the overlap between the predicted forged region and the ground truth.

4.3.2 Image-Level Metrics

These metrics evaluate whether an image is correctly classified as forged or not, regardless of the specific forged regions by a technique. Image-level metrics evaluate the performance of an algorithm in terms of classifying an entire image as either: Forged (manipulated) or Authentic (original). These metrics treat each image as a whole and are not concerned with the exact location or shape of the manipulated region. These metrics are:

(a) Binary Classification Accuracy: Proportion of images correctly classified as forged or authentic.
(b) AUC-ROC (Area Under the ROC Curve): Plots the True Positive Rate vs. False Positive Rate. A higher AUC indicates better discrimination between forged and authentic images.
(c) AP (Average Precision): The average of precision values across different thresholds. Used in cases where the algorithm gives a confidence score.

4.3.3 Region-Level Metrics

These metrics focus on the detection and localization of manipulated regions and evaluate the correctness of detected forged regions within an image, in between pixel-level (fine-grained) and image-level (coarse-grained) evaluations.

Region level metrics assess how well the detection algorithm identifies and localizes entire forged regions, rather than individual pixels or entire images and evaluate how accurately the boundary of the detected forgery matches the ground truth boundary. These metrics are:

(a) Localization Accuracy: Percentage of images where the manipulated region is correctly located.
(b) Matthews Correlation Coefficient (MCC): Considers all four confusion matrix values and provides a balanced measure even if the classes are imbalanced.
(c) Boundary-based Metrics (e.g., Boundary F1-Score)

Confusion matrix is used to evaluate the performance of a classification model, which is a table and shows the counts of correct and incorrect predictions in following form:

	Predicted as Forged image	**Predicted as Genuine image**
Forged image	True Positive	False Negative
Genuine image	False Positive	True Negative

Some metrics (Accuracy, Precision, Recall, F1-Score, Specificity, IoU, MCC) can be calculated directly derived from the confusion matrix and metrics like AUC-ROC, Average Precision are indirectly related to the confusion matrix.

4.4 Conclusion

The diversity and richness of the datasets given in this chapter provide a robust foundation for training, evaluating, and benchmarking modern algorithms aimed at detecting forged digital images and distinguishing synthetic images from real images. Selecting datasets with varied generation techniques and realistic perturbations ensures better generalization and applicability in real-world forensic tasks. Various metrics are provided which are used for evaluation of image forgery detection algorithms.

References

1. Tralic, D., Zupancic, I., Grgic, S., Grgic, M.: CoMoFoD — New database for copy-move forgery detection. In: Proceedings ELMAR-2013, Zadar, Croatia, pp. 49–54. Croatian Society Electronics in Marine (2013)
2. Christlein, V., Riess, C., Jordan, J., Angelopoulou, E.: An evaluation of popular copy-move forgery detection approaches. IEEE Trans. Inf. Forensics Secur. **7**(6), 1841–1854 (2012). https://doi.org/10.1109/TIFS.2012.2218597
3. Silva, E., Carvalho, T., Ferreira, A., Rocha, A.: Going deeper into copy-move forgery detection: exploring image telltales via multi-scale analysis and voting processes. J. Vis. Commun. Image Represent. **29**, 16–32 (2015). https://doi.org/10.1016/j.jvcir.2015.01.016

4. Amerini, I., Ballan, L., Caldelli, R., Del Bimbo, A., Serra, G.: A SIFT-based forensic method for copy–move attack detection and transformation recovery. IEEE Trans. Inf. Forensics Secur. **6**(3), 1099–1110 (2011). https://doi.org/10.1109/TIFS.2011.2129512
5. Dong, J., Wang, W., Tan, T.: CASIA image tampering detection evaluation database. In: 2013 IEEE China Summit and International Conference on Signal and Information Processing, Beijing, China, pp. 422–426. IEEE (2013). https://doi.org/10.1109/ChinaSIP.2013.6625374
6. Hossain, F., Gul, A., Raja, R., Dagiuklas, T., Galkandage, C.: Forgery image dataset. IEEE Dataport. (2022). https://doi.org/10.21227/9dmj-yn86
7. Cozzolino, D., Poggi, G., Verdoliva, L.: Efficient dense-field copy–move forgery detection. IEEE Trans. Inf. Forensics Secur. **10**(11), 2284–2297 (2015). https://doi.org/10.1109/TIFS.2015.2455334
8. https://www.ee.columbia.edu/ln/dvmm/publications/04/TRsplicingDataSetttng.pdf
9. Guan, H., et al.: MFC datasets: large-scale benchmark datasets for media forensic challenge evaluation. In: 2019 IEEE Winter Applications of Computer Vision Workshops (WACVW), Waikoloa, HI, USA, pp. 63–72. IEEE (2019). https://doi.org/10.1109/WACVW.2019.00018
10. Tokuda, E., Pedrini, H., Rocha, A.: Computer generated images vs. digital photographs: a synergetic feature and classifier combination approach. J. Vis. Commun. Image Represent. **24**(8), 1276–1292 (2013). https://doi.org/10.1016/j.jvcir.2013.08.009
11. Ng, T.-T., Chang S.-F., Hsu, J., Pepeljugoski, M.: Columbia photographic images and photorealistic computer graphics dataset. ADVENT Technical Report #205-2004-5, Columbia University (2005, February)
12. Reghioua, I., Namani, M.Y., Bendiab, G., Labiod, M.A., Shiaeles, S.: DeepGuardDB: real and text-to-image synthetic images dataset. IEEE Dataport. (2024). https://doi.org/10.21227/10ap-pk52
13. He, Y., et al.: ForgeryNet: a versatile benchmark for comprehensive forgery analysis. In: 2021 IEEE/CVF Conference on Computer Vision and Pattern Recognition (CVPR), Nashville, TN, USA, pp. 4358–4367. IEEE (2021). https://doi.org/10.1109/CVPR46437.2021.00434
14. Dang, H., Liu, F., Stehouwer, J., Liu, X., Jain, A.K.: On the detection of digital face manipulation. In: 2020 IEEE/CVF Conference on Computer Vision and Pattern Recognition (CVPR), Seattle, WA, USA, pp. 5780–5789. IEEE (2020). https://doi.org/10.1109/CVPR42600.2020.00582
15. Novozámský, A., Mahdian, B., Saic, S.: IMD2020: a large-scale annotated dataset tailored for detecting manipulated images. In: 2020 IEEE Winter Applications of Computer Vision Workshops (WACVW), Snowmass, CO, USA, pp. 71–80. IEEE (2020). https://doi.org/10.1109/WACVW50321.2020.9096940

Chapter 5
Challenges in Digital Image Forgery Detection

Abstract Image forgery detection faces growing challenges due to advanced manipulation techniques like GAN-based deepfakes and widespread post-processing. Despite progress in deep learning and transformer-based methods, issues such as poor model generalization, low-quality image handling, limited datasets, and lack of standardized evaluation hinder real-world effectiveness. This chapter highlights these key obstacles and stresses the need for robust, adaptable detection models, realistic generalized datasets, and unified evaluation protocols to ensure reliability in practical applications.

Keywords Deepfake · Forgery detection · Realistic image forgery dataset · Image manipulation

5.1 Introduction

Image forgery detection is a crucial step in digital forensics and cyber security. Due to easy availability of sophisticated image manipulation software, image forgeries whether simple copy-move forgery or deepfakes generated by generative adversarial networks (GANs) have become very common and hence ensuring the authenticity of digital images is now more difficult than ever. These image manipulations not only challenge the authenticity of visual content but also threaten the integrity of legal evidence, scientific research output, credibility of media and national security.

Advancements have been made in development of image forgery detection techniques based on deep learning, transformers, and multimodal models, but still the complex challenges in image forgery detection exist. These challenges include the lack of generalization of a technique, vulnerability of detection model, adversarial attacks increasing ethical and legal implications and unavailability of test datasets. This chapter describes these challenges, and need to be addressed.

© The Author(s), under exclusive license to Springer Nature Singapore Pte Ltd. 2026
V. Tyagi, *Digital Image Forgery Detection*, SpringerBriefs in Computer Science, https://doi.org/10.1007/978-981-95-3004-5_5

5.2 Challenges Faced in Development of an Effective Robust Image Forgery Detection Model

5.2.1 Generalization of Image Forgery Detection Techniques

In general, image forgery detection models are developed for specific types of forgery and are trained on fixed datasets, making them unusable when applied to real-world or new attacks. This lack of generalization to unseen forgeries in an image forgery detection model is a major challenge. Even high-performing models are vulnerable to adversarial attacks like a very small manipulation. These can make forged images appear genuine and bypass deep learning models.

Many deep learning models, including CNNs and transformers, operate as black boxes, making it difficult to understand their decision-making processes. This is especially problematic in sensitive applications like legal forensics.

Increasingly, forgeries are not confined to the image alone. There may be context manipulation, for instance, matching a real image with false metadata or misleading text (e.g., in cheap fakes). Techniques are required to detect inconsistencies especially in social media settings by leveraging context-aware and multimodal detection.

Modern generative models like StyleGAN3 and Stable Diffusion can create photorealistic fake images. It is a big challenge to develop generalized forgery detection model that can detect synthetic images generated by such algorithms.

5.2.2 Low Quality Images and Post-processing Operations

The efficacy of forgery detection algorithms is significantly reduced when images are of low quality or have undergone post-processing operation.

Image quality degradation can be due to: sensor noise, lossy compression of image, resizing, transmission errors, or intentional tampering. Lossy image compression techniques, such as JPEG, reduce file size by removing image information, affecting the clues that can help in detecting authenticity of image. JPEG compression introduces blocking artifacts, which may either hide actual tampering traces or be mistakenly interpreted as forgeries.

Forgers often apply post-processing techniques like recompression, resizing, noise addition, blurring to hide manipulation traces. Recompression is a common issue when an already compressed image is compressed again, introducing additional artifacts and noise. This is very common in social media platforms (e.g., Facebook, WhatsApp), where images are automatically recompressed before uploading or sharing through these platforms. For example, copy-move forgeries can be concealed by blurring borders and then compressing the image. This makes even patch-based matching approaches like block-matching or SIFT ineffective in detecting cloned regions.

These processes obscure forgery clues or introduce misleading artifacts, making it more challenging to distinguish authentic manipulations from compression effects.

These modifications obscure forensic features such as inconsistencies in noise patterns, JPEG artifacts, or edge irregularities. These operations not only degrade image quality but also distort pixel-level cues used in detection. Handling of such post-processing operations is a big challenge in developing an effective forgery detection technique.

However, most datasets do not include such versions, meaning models fail to handle common distortions like JPEG artifacts, downsampling, or color-space conversion.

The NIST Media Forensics Challenge datasets and FaceForensics++ [1] have attempted to simulate compression variants, but such datasets are still inadequate and domain-specific (e.g., only face forgeries) [2].

Deep learning-based forgery detection models (e.g., CNNs, transformers) perform well on high-quality training datasets but often fail when tested on low-resolution or recompressed images. These models are highly sensitive to the input distribution, and fail when images differ from the training data.

This generalization gap implies that models trained on clean images often misclassify noisy or compressed images, leading to poor performance in real-world scenarios.

5.2.3 *Feature Representation and Invariance*

Selection of robust and discriminative features is very important in the success of an image forgery detection technique. Some features may work very well but performance may degrade in presence of some operations. For example, keypoint-based features (e.g., SIFT, SURF) and moment-based features (e.g., Gaussian-Hermite Moments) are used in a number of image forgery detection techniques, they often fail under scale, rotation, or illumination changes.

Similarly, traditional hand-crafted features, such as DCT coefficients or color moments, generally do not work under quality degradation. While deep features offer better generalization, they too require adaptation. Models like XceptionNet have shown success in detecting deepfakes but still struggle with compressed variants due to loss of discriminatory features [1].

Hybrid approaches that combine both handcrafted and learned features are still under development and often lack robustness across varying image qualities.

5.2.4 Image Forgery Detection Dataset Limitations

State-of-the-art image forgery detection techniques rely on image datasets for training, validation, and evaluation. However, lack of availability, quality, and diversity of ideal datasets used for training, validation and testing the models have introduced several limitations and performance bottlenecks. These issues are not only technical but also methodological, as they influence the generalization ability and real-world applicability of forgery detection system.

Although a number of datasets are available for training and evaluation of image forgery detection techniques, but available benchmark datasets lack diversity in terms of camera models, image types, compression artifacts, and manipulation complexity. Many datasets are either synthetic or overly clean and do not reflect the real-world complexity of image forgeries. This leads to overfitting and poor generalization.

Most existing datasets such as CASIA [3], CoMoFoD [4], and Columbia [5] involve synthetic manipulations e.g., copy-move or splicing, that are generated in controlled settings. These forgeries often lack the complexity and variability found in real-world tampered images.

This synthetic nature causes a distributional gap between training and test data in practical scenarios, leading to poor generalization. For example, a model trained on the CASIA dataset may achieve high accuracy within the dataset but fail to detect forgeries on social media images, where multiple compression and post-processing operations are involved [6].

These datasets do not adequately simulate the low-quality and recompressed images common in real applications. As a result, detection algorithms evaluated on these datasets may not perform well on real time data. Many datasets are biased toward certain forgery types (e.g., copy-move), image formats (e.g., JPEG only), or scene categories (e.g., indoor scenes). This lack of diversity limits a model's ability to generalize across various manipulation styles and domains.

Moreover, forged regions in many datasets are larger and more obvious than in real forgeries, leading to a bias in feature learning. For example, convolutional neural networks may learn to focus on texture or boundary changes that do not actually occur in subtle or semantically meaningful forgeries [7].

Many of the available datasets are small in size, with only a few hundred manipulated images. Deep learning models, which typically require thousands or millions of samples, often overfit to these limited datasets, learning dataset-specific artifacts rather than general forgery features.

Annotation quality is another concern. Ground truth masks may be imprecise or unavailable, which is problematic for pixel-level localization tasks. Datasets such as CoMoFoD provide masks, but inconsistencies in resolution or alignment make it difficult to use them directly in training pipelines [8].

Almost all datasets are single-frame and image-only, lacking video sequences or metadata such as GPS, EXIF, or device information. Forgery detection in multimedia contexts (e.g., social media posts) requires temporal analysis or cross-modal verification.

The lack of datasets with multi-modal or temporal data severely limits the development of hybrid forensic tools that combine visual, textual, and sensor-based clues [9].

Another major issue is the absence of standardized protocols for evaluation across datasets. Different studies use different training-test splits, quality settings, and manipulation types, making it difficult to compare models fairly. For instance, one model might report 98% accuracy on CASIA with a 50–50 split, while another might use an 80–20 split or a different dataset entirely, invalidating comparative claims. Additionally, there is no consensus on whether to include benign post-processing (like blurring or resizing) in test images [10].

Due to the above limitations, many models demonstrate high performance in closed-set evaluations but fail in open-set, cross-dataset tests. This causes a false sense of progress in the field and leads to the development of over-engineered models that fail in practical deployments.

Moreover, academic research often emphasizes novelty in architecture over real-world applicability, partly due to the lack of suitable benchmarks that simulate deployment conditions [6].

To overcome dataset-induced performance challenges, the following need to be considered:

- Realistic Dataset Generation: Creating datasets with real forgeries collected from social media or forensic investigations.
- Compression and Noise Variants: Including multiple versions of the same image under varying compression, resolution, and noise levels.
- Cross-Domain Benchmarks: Evaluating models across multiple datasets with standardized splits and metrics.
- Multi-Modal Datasets: Including metadata (e.g., EXIF, GPS) and multiple image types (e.g., screenshots, camera captures).
- Crowdsourced or Synthetic Data Augmentation: Leveraging GANs or synthetic pipelines to augment underrepresented forgery types.

5.2.5 Precision in Localization of Tampered Regions

Accurate localization of tampered regions in a forged image is very important, especially in scientific research, legal and journalistic domains. However, identifying tampered regions with pixel-level accuracy remains a challenging task for most image forgery detection models. Image forgery detection algorithms either fail to mark boundaries precisely or produce high false positives. Therefore there is need to develop forgery detection techniques that can precisely mark tampered regions in a forged image [11].

5.2.6 Absence of Standardized Evaluation Protocols

There is no universally accepted evaluation protocol that tests image forgery detection systems under a range of image qualities and compression levels. As a result, research outputs are hard to compare, and many methods report high accuracy under controlled conditions but fail in real-world deployment. Another issue of non-standardization is use of different evaluation metrics (accuracy, F1-score, AUC, pixel-wise IoU) in different research, making direct comparison difficult.

A standardized benchmark that includes multi-resolution, recompressed, and multi-format (e.g. JPEG, PNG, WebP) images and a standard set of evaluation matrices is essential for evaluating robustness.

5.2.7 Legal and Ethical Issues

The rapid evolution of deepfake and image manipulation technologies has significantly outpaced the development of corresponding legal frameworks and regulatory mechanisms. Despite considerable advances in detection algorithms, ranging from traditional techniques to modern deep learning-based models, there remains a conspicuous absence of universal legal standards for defining, regulating, or penalizing visual forgeries.

From legal point of view, jurisdictions around the world differ significantly in handling digital image forgery. Some countries have introduced laws targeting specific use cases, such as revenge pornography or election interference, but these are often reactive and lack a cohesive, technology-neutral framework. For instance, while the United States has passed state-level legislation against deepfakes in political campaigns and non-consensual pornography, there is no comprehensive federal statute that addresses the broader spectrum of visual misinformation [12]. Similarly, the European Union's General Data Protection Regulation (GDPR) provides some legal recourse for individuals whose likenesses are misused, but it does not explicitly address synthetic media or manipulated content.

The ethical issues are also equally important. The use of powerful generative tools (e.g., GANs, diffusion models, and face-swapping software) raises questions about informed consent, authenticity, and the weaponization of media. Manipulated images can be used to defame individuals, incite violence, or erode public trust in media institutions [13]. The ethical responsibility thus extends not only to creators and disseminators of such content but also to platform providers, researchers, and policy-makers. Researchers, in particular, face a dual challenge: advancing detection capabilities while ensuring that published methodologies are not misused to improve evasion techniques.

Moreover, there is a growing need for standards of accountability, both technical and legal. These include chain-of-custody protocols, watermarking or provenance metadata standards, and certification mechanisms for authentic digital content [14].

Interdisciplinary collaboration between computer scientists, legal scholars, ethicists, and policy-makers is vital to build an ecosystem where technological solutions are embedded within a robust legal and ethical scaffold.

5.3 Conclusion

Despite the progress made in the field, image forgery detection remains a challenging task due to multiple factors, including diverse manipulation techniques, computational trade-offs, limited real-world datasets, and the evolving sophistication of digital forgeries. There is a need of developing generalized and robust techniques that can perform reliably across multiple domains and under real-world constraints.

The advancement of image forgery detection is heavily contingent upon the quality, realism, and diversity of datasets used in developing a forgery detection model. Current limitations, ranging from fine manipulations to poor annotations and homogeneity, are big challenges in the development of generalizable and robust detection models. Addressing these dataset challenges is required for meaningful progress in both academic and applied image forensics.

The detection of image forgeries on low-quality and recompressed images is another important challenge. Loss of forensic cues, compression artifacts, poor generalization of models, and the lack of realistic datasets collectively hamper performance. Future research must prioritize robustness, dataset realism, and adaptability of models to ensure real-world efficacy.

The field of image forgery detection is a critical intersection of technology, policy, and ethics. While developments have been made in image forgery detection techniques, using machine learning and deep neural networks, these systems remain vulnerable to generalization issues, adversarial attacks, and lack of interpretability. Latest techniques such as causal inference, multimodal analysis, and transformer-based models offer promising directions. However, to maintain the authenticity and secure the integrity of digital images, a holistic approach that integrates robust detection models, dynamic datasets, legal frameworks, and ethical supervision is required.

References

1. Rossler, A., Cozzolino, D., Verdoliva, L., Riess, C., Thies, J., Nießner, M.: Faceforensics++: learning to detect manipulated facial images. In: Proceedings of the IEEE/CVF International Conference on Computer Vision, Seoul, 27–28 October 2019, pp. 1–11. IEEE (2019). https://doi.org/10.1109/ICCV.2019.00009
2. Amerini, I., Ballan, L., Caldelli, R., Del Bimbo, A., Del Tongo, L., Serra, G.: Copy-move forgery detection and localization by means of robust clustering with J-Linkage. Signal Process. Image Commun. **28**(6), 659–669 (2013). https://doi.org/10.1016/j.image.2013.03.006J

3. Dong, J., Wang, W., Tan, T.: CASIA image tampering detection evaluation database. In: 2013 IEEE China Summit and International Conference on Signal and Information Processing, Beijing, China, pp. 422–426. IEEE (2013). https://doi.org/10.1109/ChinaSIP.2013.6625374
4. Tralic, D., Zupancic, I., Grgic, S., Grgic, M.: CoMoFoD — New Database for Copy-Move Forgery Detection. In: Proceedings ELMAR-2013, Zadar, Croatia, pp. 49–54. Croatian Society Electronics in Marine (2013)
5. Ng, T.T., Chang, S., Sun, Q.: A data set of authentic and spliced image blocks. ADVENT Technical Report 203-2004-3, Columbia University (2004, June)
6. Verdoliva, L.: Media forensics and deepfakes: an overview. IEEE J. Sel. Top. Signal Process. **14**(5), 910–932 (2020). https://doi.org/10.1109/JSTSP.2020.3002101
7. Zhou, P., Han, X., Morariu, V.I., Davis, L.S.: Learning rich features for image manipulation detection. In: 2018 IEEE/CVF Conference on Computer Vision and Pattern Recognition, Salt Lake City, UT, USA, pp. 1053–1061. IEEE (2018). https://doi.org/10.1109/CVPR.2018.00116
8. Bappy, J.H., Roy-Chowdhury, A.K., Bunk, J., Nataraj, L., Manjunath, B.S.: Exploiting spatial structure for localizing manipulated image regions. In: 2017 IEEE International Conference on Computer Vision (ICCV), Venice, Italy, pp. 4980–4989. IEEE (2017). https://doi.org/10.1109/ICCV.2017.532
9. Barni, M., Stamm, M., Tondi, B.: Adversarial multimedia forensics: overview and challenges ahead. In: 2018 26th European Signal Processing Conference (EUSIPCO), pp. 962–966 (2018). https://doi.org/10.23919/EUSIPCO.2018.8553305
10. Christlein, V., Riess, C., Jordan, J., Riess, C., Angelopoulou, E.: An evaluation of popular copy-move forgery detection approaches. IEEE Trans. Inf. Forensic Secur. **7**(6), 1841–1854 (2012). https://doi.org/10.1109/TIFS.2012.2218597
11. Meena, K.B., Tyagi, V.: Image forgery detection: survey and future directions. In: Data, Engineering and Applications. Springer, Singapore (2019). https://doi.org/10.1007/978-981-13-6351-1_14
12. Chesney, S., Citron, D.: Deep fakes: a looming challenge for privacy, democracy, and national security. Calif. Law Rev. **107**(6), 1753–1819 (2019). https://doi.org/10.2139/ssrn.3213954
13. Westerlund, M.: The emergence of deepfake technology: a review. Technol. Innov. Manag. Rev. **9**(11), 40–53 (2019). https://doi.org/10.22215/timreview/1282
14. Farid, R.: Creating, verifying and distributing trusted digital content. J. Online Trust Saf. **1**(1), 1–13 (2021) https://cyber.fsi.stanford.edu/publication/creating-verifying-and-distributing-trusted-digital-content

Chapter 6
Applications of Digital Image Forgery Detection

Abstract This chapter explores key applications of digital image forgery detection across journalism, law enforcement, social media, and academic research. With the widespread use of manipulated images for misinformation, legal deception, and academic misconduct, forgery detection has become vital. Techniques such as deep learning, noise analysis, and metadata verification are now being integrated into media verification, forensic investigations, content moderation, and research integrity checks. Despite ongoing challenges such as legal gaps and technical limitations; image forgery detection is emerging as a crucial tool to uphold truth, justice, and public trust in the digital age.

Keywords Image forgery detection · Journalism · Misinformation · Image forensics · Research integrity

6.1 Introduction

Digital images have emerged as one of the most powerful communication tools in the digital age. From social media platforms to mainstream journalism and in research, digital images and digital visuals are critical in shaping public opinion, informing legal decisions, and communicating research results. However, digital image manipulation software has given rise to the challenge of image forgery. Image forgery detection, powered by advances in computer vision and deep learning, has emerged as an essential technological defense against digital misinformation and deception. Its applications span many domains. This chapter focuses on three major areas: journalism and media, law enforcement and legal frameworks, and social media and content moderation.

© The Author(s), under exclusive license to Springer Nature Singapore Pte Ltd. 2026
V. Tyagi, *Digital Image Forgery Detection*, SpringerBriefs in Computer Science, https://doi.org/10.1007/978-981-95-3004-5_6

6.2 Applications in Journalism and Media

Journalism relies heavily on the trust of its audience. Visual evidence such as photographs and videos lend credibility to news stories. However, the advent of sophisticated photo editing tools has made it possible to manipulate images in ways that can significantly alter the context of news. Such manipulations may be used to sensationalize reports, mislead readers by using forged photographs or videos. Presenting a forged image or video as an authentic one, has many consequences such as:

- Incite panic or violence, especially in politically or culturally sensitive regions
- Damage the reputation of media outlets
- Mislead the public and influence electoral or policy decisions

An example is the manipulation of war images to exaggerate destruction or misattribute blame, which has occurred in various geopolitical contexts.

Application of Image forgery detection in confirming authenticity of an image is to embed it into newsroom verification pipeline to authenticate incoming images. Techniques like Error Level Analysis (ELA), noise analysis, copy-move forgery detection, and deep learning-based classifiers can be used to flag suspicious content.

Some leading agencies now collaborate with AI developers to integrate forgery detection tools into their verification processes. These systems often work together with human fact-checkers.

Advanced forgery detection systems are not only identify pixel-level image manipulations; they also assess semantic consistency between image and related data. For example, if a caption in a photo claims that the photo was taken in New Delhi but the image metadata or background suggests otherwise, advanced AI fast checkers can flag such inconsistencies.

6.3 Applications in Law Enforcement and Legal Frameworks

Images and videos are very important in the area of digital forensics. In court of law, digital images can be used as forensic evidence in complex cases. These images may be related to crime scene, CCTV footage, mobile phone captures, or a social media post. Forged images can mislead investigations or be used to wrongly punish innocent individuals. Therefore it is very essential to detect forged images so that justice can be served. Forged images can mislead legal and forensic investigation as such images can be used to:

- Hide identities in surveillance
- Create false evidence in court
- Alter photos to support fraudulent claims like insurance claims

Forgery detection techniques help forensic experts in validating the authenticity of images. Methods such as JPEG artifact analysis, Photo Response Non-Uniformity (PRNU), and noise pattern comparison are accepted in digital forensic labs for the same.

Recent advances in Convolutional Neural Networks (CNNs) and Transformer-based models can help forensic analysts to detect highly undetectable tampering that cannot be detected through classical forgery detection methods.

Despite developments of many efficient techniques in this area, there are still open challenges like:

- No proper standard operating procedures (SOPs) for image forgery detection in many police departments
- Absence of specific cybercrime laws in many countries that can be applied to criminalize digital image forgery
- Delays in accepting algorithmic outputs as decisive evidence due to complexity of the system

There is an urgent need of court laws that defines what constitutes digital forgery and way to prosecute such offenses.

6.4 Applications in Social Media and Content Moderation

Social media platforms like Facebook, X (formerly Twitter), Instagram, WhatsApp, Snapchat, YouTube host billions of images and videos newly generated uploads everyday. While this supports content creation, but it has become an effective platform for miscreants to propagate image-based misinformation, harassment, and hate speech through forged images.

Manipulated images having misinformation spread quickly, often before fact-checkers can detect these. This may result in dangerous consequences, such as mob violence in response to forged photos.

Social media platforms are now integrating AI-driven forgery detection systems to assist in content moderation using fact checkers. These tools can detect possibility of:

- Morphed images used in cyberbullying or revenge porn
- Cloned or doctored images used in scams and fraud
- Deepfake videos impersonating celebrities or politicians

Facebook's Deepfake Detection Challenge (DFDC), initiated in 2019, was a key step toward developing and evaluating scalable solutions. Similarly, other social media platforms are using machine learning classifiers to flag manipulated media, often in real time.

While image forgery detection AI tools can filter out the bulk of forged content, human moderators are often needed to assess context, intent, and potential harm. This needs a multi-layered approach with following steps:

- Automated flagging of suspected forged content
- Human review for final decision-making
- Transparency to users with labels such as "Manipulated Media" or "Modified Photo"

Social media platform X for instance, began labeling tweets with fake media following incidents of political misinformation.

Social media platforms are increasingly being held accountable for misinformation. Regulatory efforts such as the Digital Services Act (EU) and the Information Technology Act, 2000 (IT Act 2000) compel platforms to detect and act upon harmful forged media. Compliance requires robust forgery detection capabilities.

Moreover, NGOs and independent fact-checkers use open-source tools like FotoForensics to aid grassroots efforts in identifying manipulated content.

6.5 Applications in Research

With the ease in availability of sophisticated image editing software, digital image manipulation has increased in scientific and academic research. Therefore for maintaining the integrity of scientific and academic research image forgery detection is required. As research outcomes relies on visual data e.g. graphs, microscopy images, satellite images, and clinical photographs—the potential for both intentional and unintentional manipulation poses a serious threat to research results credibility.

Many high-impact research journals and funding bodies now ensure that authors of research manuscripts disclose any image editing process used and adhere to strict ethical guidelines. However, image manipulation cases are increasing, especially in biomedical sciences, where duplicated or altered images, cell microscopy images, or experimental results have led to manuscript retractions. Forgery detection tools help publishers in identifying signs of data tampering before publication, ensuring that published results are based on actual evidences.

Research in the area of image forensics itself has become a rapidly expanding sub-field of computer vision and artificial intelligence. Researchers are actively developing new models, from handcrafted feature-based forgery detection systems to deep neural networks, aimed at detecting various types of image manipulations.

Interdisciplinary research collaborations can be made in the area of image forgery detection. Computer scientists, forensic analysts, legal professionals, and media experts are increasingly collaborating together to address challenges in detecting and responding to manipulated visual content.

6.6 Conclusion

Image forgery detection has become a necessity of society and research community today. As digital images' use is increasing in digital communication, the authenticity of visual content is very important to establish truth, provide justice, and build trust.

Maintaining the credibility of journalism, supporting judiciary in ensuring fair trials, helping social media platforms removing misinformation and supporting in identifying wrong information in research, all important areas can use forgery detection tools as a supporting tool. Although technical, legal, and ethical challenges exists but collaborative efforts across domains offer a potential way forward.

Chapter 7
Future Directions in Digital Image Forgery Detection Research

Abstract In this Chapter, various open research issues in the area of digital image forgery detection are discussed. Although some research work has been done on these issues, but still research is required to get effective solution to problem of image forgery detection.

Keywords Explainable AI · Few shot learning · Privacy preserving forensics

7.1 Introduction

The rapid evolution of digital content creation tools, including AI-generated media and sophisticated editing software, has made digital image forgery both easier and harder to detect. Available free or low-cost image editing tools can manipulate images and these manipulations are hard to detect by human eye. Therefore there is a strong need of developing advanced image forgery detection methods [1]. While deep learning based methods have improved forgery detection accuracy, image forgeries are becoming more dangerous and multimodal. In this chapter, future research directions are explored in the area of image forgery detection that will help in development of effective image forgery detection models.

7.2 Robust and Explainable Deep Learning Models

Deep learning models have helped significantly in increasing the efficacy of image forgery detection systems due to their powerful feature extraction capabilities. However, two important concerns of such models are: the robustness under diverse and adversarial conditions, and lack of interpretability. As the implications of image forgery affect important areas such as journalism, law agencies, and digital forensics, it becomes essential to develop image forgery detection models that are both robust and explainable. While robustness of a forgery detection system ensures

© The Author(s), under exclusive license to Springer Nature Singapore Pte Ltd. 2026
V. Tyagi, *Digital Image Forgery Detection*, SpringerBriefs in Computer Science, https://doi.org/10.1007/978-981-95-3004-5_7

reliability, explainability of the model is essential for building trust and usability in forensic applications.

Existing deep learning based forgery detection models are like black-box models that lack transparency [2]. Many existing detection methods also depend on manual human interpretation, urging fully automated and interpretable detectors. Future research should focus on interpretable AI that can use techniques like Grad-CAM, LIME, and SHAP [3]. Additionally, forgery detection models must be robust against adversarial attacks and optimized for edge deployment using efficient architectures like TinyViT [4].

Such forgery detection models will not only improve detection accuracy across varied and adversarial conditions but will also promote trust and transparency in decision-making, making this an important research topic in the area of digital image forensics.

7.3 Multimodal and Cross-Modal Forgery Detection

Digital media includes not only images, but accompanying text, metadata, audio, or video. Traditional forgery detection methods works mainly on image and don't use associated information. Combining image, text, audio, and metadata is a promising future research direction to detect inconsistencies to help in identifying forgery. Techniques such as CLIP or BLIP enable cross-modal learning, while metadata (like EXIF) and social context can enhance forgery detection [5].

In contrast, cross-modal forgery detection focuses on detecting discrepancies between modalities, such as a mismatch between the image content and its metadata e.g. location or timestamp. These approaches often use separate encoders per modality and compare their outputs for alignment or verification. Although recent advances in deep learning have enabled effective integration of multimodal information, improving detection performance, especially in social media contexts. However, challenges remain, including modality imbalance, semantic misalignment, and the inadequacy of rich multimodal forgery datasets.

7.4 Self-Supervised and Few-Shot Learning

Data inadequacy, especially of annotated forgeries, limits efficiency of deep learning based models. There is a need for automated methods less reliant on manual ground truth, such as self-supervised or few-shot approaches [6]. Self-supervised methods like SimCLR and BYOL can learn from unlabelled data, while few-shot models can adapt to new forgery types with minimal supervision. Research in the area of developing efficient image forgery detection techniques learning with minimal or unlabelled data is a need of time.

7.5 Generalization to Unknown or Emerging Forgeries

A major challenge in image forgery detection lies in building models that can generalize beyond known forgery types to detect unknown forgeries. Most image forgery detectors are trained on specific datasets containing well-defined forgery types, such as splicing, copy-move, or GAN-generated images. However, in real-world scenarios, new types of image manipulation may appear that deviate significantly from those used during development the system. Future research should focus on domain generalization, adaptive systems and continual learning to detect unseen or evolving forgery. This involves training models that are robust across different datasets and can adapt over time [7].

7.6 Spatiotemporal Forgery Localization

Pixel-level tampering detection and temporal consistency analysis in videos are key challenges. Vision Transformers (ViTs) and DETR-based models show promise for fine-grained forgery localization [8]. There is a need of developing forgery detection systems that can accurately identify where and when the manipulation occurs in a visual sequence.

7.7 Dataset Expansion and Realistic Benchmarking

Most publicly available datasets are either synthetically generated or cover a narrow range of manipulation types, sources, and conditions. This creates a dataset bias, leading models to perform well on benchmark datasets but poorly on real-world, unseen forgeries. There is a need of development of larger and more diverse datasets with pixel-level annotations. Synthetic data generation using GANs and more comprehensive evaluation metrics will enable more realistic benchmarking [9].

In addition, there is a need of more realistic benchmarking protocols. Current evaluation practices often assume well-annotated, balanced datasets and focus on binary classification or segmentation performance. However, realistic benchmarking should also assess generalization to unseen manipulations, cross-dataset performance, temporal stability, and robustness to content-level variability. Several recent initiatives have attempted to address these issues by offering larger and more varied datasets. However, many still lack high-quality pixel-level ground truth, support for multimodal content, or realistic user-driven manipulations.

Going forward, there is a need of creation of benchmark suites that combine diverse datasets, cover a wide range of forgery types, and promote standardized, open evaluation protocols. This will enable the development of more robust, generalizable, and trustworthy image forgery detection systems.

7.8 Federated and Privacy-Preserving Forensics

There is growing concern over the privacy and security of user images, especially in sensitive domains such as law enforcement, journalism, and personal media. Traditional forgery detection models requires large amounts of images to be collected and stored on shared devices, raising significant privacy risks and potential regulatory challenges under frameworks.

Federated and privacy-preserving learning approaches that allows training on decentralized data, enhancing privacy represent a promising direction for scalable and ethical forensic analysis. Privacy-preserving AI techniques such as homomorphic encryption can also be explored [10].

7.9 Real-Time Deployment

Most existing models are developed and evaluated under controlled conditions using well-curated datasets. However, actual deployment scenarios involve diverse content sources, unpredictable manipulations, varying device conditions, and real-time processing constraints, all of which impact model effectiveness. Deploying real-time detection on browsers, mobile apps, or cloud platforms requires lightweight, efficient models. Collaborative human-AI systems can further enhance reliability [11]. Moving image forgery detection system from lab to real life application requires research in model efficiency, robustness, adaptability, and usability.

7.10 Integration with Other Digital Forensics Disciplines

Image forgery detection is increasingly being integrated with other digital forensics fields such as video, audio, metadata, cloud forensics (for analyzing data stored in the cloud), and social network forensics (for understanding the spread of forged images on online platforms) and network forensics, to enable more comprehensive and reliable analysis. In real-world scenarios, manipulated content often exists alongside other digital evidence, making cross-domain correlation essential. For example, detecting image tampering can be strengthened by verifying metadata, analyzing audio-visual synchronization in videos, or matching content with network logs.

Such integration improves detection accuracy, supports context-aware interpretation, and enhances legal admissibility. Research in this integration is another open problem in development of an efficient image forgery detection system.

7.11 Legal and Ethical Considerations

As image forgery detection technologies advance, it is crucial to address the associated legal and ethical implications. From a legal perspective, forgery detection tools must meet standards for evidence admissibility and data integrity, especially when used in judicial or investigative contexts. The outputs of detection models should be explainable, reproducible, and verifiable to ensure their acceptance in court or regulatory settings.

Ethically, the deployment of such tools must balance security and privacy. Over-reliance on automated detection without human oversight risks false accusations, especially when models are applied to content from diverse populations or manipulated datasets. Additionally, models trained on personal or sensitive data raise concerns regarding informed consent, data protection, and bias.

Future detection systems must be explainable and legally admissible and detection systems must align with policy frameworks [12, 13].

7.12 Increasing Public Awareness and Education

As image forgeries become more sophisticated and accessible, raising public awareness and education is essential to mitigate their societal impact. Even the most advanced detection tools are ineffective if end-users such as journalists, educators, law enforcement, or everyday social media users are not able to know the output of a detection system. Therefore, raising public awareness and education is essential to mitigate their societal impact. Therefore, building media literacy and promoting critical evaluation skills is a key line of defense against misinformation and visual deception. Ultimately, the success of forgery detection efforts depends not just on technical solutions, but on an informed and vigilant public that understands both the risks of manipulated media and the tools available to combat it.

By pursuing these future directions, researchers can contribute to a more robust and effective defense against the growing problem of image forgery and its potential negative consequences in various domains.

7.13 Conclusion

The future of effective image forgery detection research lies in the fusion of robust, explainable, and generalizable AI models, enhanced by multi-modal data and realistic conditions. Open benchmarks, cross-disciplinary approaches, and focus on ethical deployment are required. Collaboration among AI researchers, security experts, policymakers, and industry is must to make digital image forgery detection efficient. Building media literacy and promoting critical evaluation skills is also a key line of defense against misinformation and visual deception.

References

1. Meena, K.B., Tyagi, V.: Image forgery detection: survey and future directions. In: Data, Engineering and Applications, pp. 163–194. Springer, Singapore (2019). https://doi.org/10.1007/978-981-13-6351-1_14
2. Solanke, A.A.: Explainable digital forensics AI: towards mitigating distrust in AI-based digital forensics analysis using interpretable models. Forensic Sci. Int. Digit. Investig. **42**(Supplement), 301403 (2022). https://doi.org/10.1016/j.fsidi.2022.301403
3. Panati, C., Wagner, S., Brüggenwirth, S.: Feature relevance evaluation using Grad-CAM, LIME and SHAP for deep learning SAR data classification. In: 2022 23rd International Radar Symposium (IRS), Gdansk, Poland, pp. 457–462 (2022). https://doi.org/10.23919/IRS54158.2022.9904989
4. Wu, K., Zhang, J., Peng, H., Liu, M., Xiao, B., Jianlong, F., Yuan, L.: TinyViT: fast pretraining distillation for small vision transformers. In: Computer Vision – ECCV 2022: 17th European Conference, Tel Aviv, Israel, October 23–27, 2022, Proceedings, Part XXI, pp. 68–85. Springer-Verlag, Berlin, Heidelberg (2022). https://doi.org/10.1007/978-3-031-19803-8_5
5. Barni, M.: Multimodal forensics: the next frontier. EURASIP J. Inf. Secur. (2022). https://doi.org/10.1186/s13635-022-00113-x
6. Uelwer, T., Robine, J., Wagner, S.S., et al.: A survey on self-supervised methods for visual representation learning. Mach. Learn. **114**, 111 (2025). https://doi.org/10.1007/s10994-024-06708-7
7. Guan, W., Wang, W., Dong, J., Peng, B.: Improving generalization of deepfake detectors by imposing gradient regularization. IEEE Trans. Inf. Forensics Secur. **19**, 5345–5356 (2024). https://doi.org/10.1109/TIFS.2024.3396064
8. Atak, I.G., Yasar, A.: Image forgery detection by combining Visual Transformer with Variational Autoencoder Network. Appl. Soft Comput. **165**, 112068 (2024). https://doi.org/10.1016/j.asoc.2024.112068
9. Cardenuto, J.P., Rocha, A.: Benchmarking scientific image forgery detectors. Sci. Eng. Ethics. **28**, 35 (2022). https://doi.org/10.1007/s11948-022-00391-4
10. Iyengar, S.S., Nabavirazavi, S., Hariprasad, Y., Prasad, H.B., Mohan, C.K.: Privacy-preserving AI (federated learning) for digital forensics. In: Artificial Intelligence in Practice. Signals and Communication Technology. Springer, Cham (2025). https://doi.org/10.1007/978-3-031-89327-8_5
11. Lu, Y., Ebrahimi, T.: Assessment framework for deepfake detection in real-world situations. EURASIP J. Image Video Process. **2024**, 6 (2024). https://doi.org/10.1186/s13640-024-00621-8
12. Billah, M.: Developing an explainable AI system for digital forensics: enhancing trust and transparency in flagging events for legal evidence. J Forensic Sci. Res. **9**(2), 109–116 (2025). https://doi.org/10.29328/journal.jfsr.1001089
13. Rocha, A., Scheirer, W., Boult, T., Goldenstein, S.: Vision of the unseen: current trends and challenges in digital image and video forensics. ACM Comput. Surv. **43**(4), Article 26 (2011). https://doi.org/10.1145/1978802.1978805

Bibliography

1. Abdalla, Y., Tariq Iqbal, M., Shehata, M.: Copy-move forgery detection and localization using a generative adversarial network and convolutional neural-network. Information (Switzerland). **10**(9) (2019). https://doi.org/10.3390/info10090286
2. Abidin, A.B.Z., Majid, H.B.A., Samah, A.B.A., Hashim, H.B.: Copy-move image forgery detection using deep learning methods: a review. In: 2019 6th International Conference on Research and Innovation in Information Systems (ICRIIS), pp. 1–6. IEEE (2019). https://doi.org/10.1109/ICRIIS48246.2019.9073569
3. Abou-Zbiba, W., Benbrahim, H., Benhaddou, D., El Bakkali, H.: Is attention mechanism enough for accurate deep learning-based image forgery detection models? In: 2025 International Wireless Communications and Mobile Computing (IWCMC), Abu Dhabi, United Arab Emirates, pp. 1229–1234. IEEE (2025). https://doi.org/10.1109/IWCMC65282.2025.11059610
4. Abraham, A.R., Rahim, M.S.M., Sulong, G.: Bin: splicing image forgery identification based on artificial neural network approach and texture features. Clust. Comput. **22**, 647–660 (2019)
5. Achanta, R., Shaji, A., Smith, K., Lucchi, A., Fua, P., Süsstrunk, S.: Slic superpixels. Technical report, EPFL (2010)
6. Agarwal, R., Kashyap, A., Yadav, H., Mathur, H., Rajesh, B., Javed, M.: Triple-stream SWIN transformer based encoder-decoder for image forgery localization. In: 2025 8th International Conference on Information and Computer Technologies (ICICT), Hawaii-Hilo, HI, USA, pp. 492–497. IEEE (2025). https://doi.org/10.1109/ICICT64582.2025.00083
7. Agarwal, R., Verma, O.: An efficient copy move forgery detection using deep learning feature extraction and matching algorithm. Multimed. Tools Appl. **79** (2020). https://doi.org/10.1007/s11042-019-08495-z
8. Agarwal, R., Verma, O.P.: Robust copy-move forgery detection using modified superpixel based FCM clustering with emperor penguin optimization and block feature matching. Evol. Syst. **13**(1), 27–41 (2022). https://doi.org/10.1007/s12530-021-09367-4
9. Ahmed, B., Gulliver, T.A., alZahir, S.: Blind copy-move forgery detection using SVD and KS test. SN Appl. Sci. **2**(8) (2020). https://doi.org/10.1007/s42452-020-3181-6
10. Ahmed, B., Gulliver, T.A., alZahir, S.: Image splicing detection using mask-RCNN. SIViP. **14**, 1035–1042 (2020). https://doi.org/10.1007/s11760-020-01636-0
11. Akshara, A., Gayathri, R., Alex Raj, S.M.: Leveraging DeepLabV3 for accurate localization of image forgeries. In: 2025 6th International Conference on Control, Communication and Computing (ICCC), Thiruvanathapuram, India, pp. 1–5. IEEE (2025). https://doi.org/10.1109/ICCC64910.2025.11077204

12. Akshara, A., Gayathri, R., Alex Raj, S.M.: Localized forgery detection: integrating DeepLabV3 with error level analysis. In: 2025 3rd International Conference on Intelligent Systems, Advanced Computing and Communication (ISACC), Silchar, India, pp. 458–463. IEEE (2025). https://doi.org/10.1109/ISACC65211.2025.10969257
13. Al_azrak, F.M., Elsharkawy, Z.F., Elkorany, A.S., El Banby, G.M., Dessowky, M.I., El-Samie, F.E.A.: Copy-move forgery detection based on discrete and SURF transforms. Wirel. Pers. Commun. **110**(1), 503–530 (2020). https://doi.org/10.1007/s11277-019-06739-7
14. Alberry, H.A., Hegazy, A.A., Salama, G.I.: A fast SIFT based method for copy move forgery detection. Future Comput. Inform. J. **3**(2), 159–165 (2018). https://doi.org/10.1016/j.fcij.2018.03.001
15. Alencar, A.L., Lopes, M.D., Fernandes, A.M.R., Anjos, J.C.S., De Paz Santana, J.F., Leithardt, V.R.Q.: Detection of forged images using a combination of passive methods based on neural networks. Future Internet. **16**(3), 97 (2024). https://doi.org/10.3390/fi16030097
16. Alkawaz, M.H., Sulong, G., Saba, T., Rehman, A.: Detection of copy-move image forgery based on discrete cosine transform. Neural Comput. & Applic. **30**(1), 183–192 (2018). https://doi.org/10.1007/s00521-016-2663-3
17. Al-Qershi, O.M., Khoo, B.E.: Evaluation of copy-move forgery detection: datasets and evaluation metrics. Multimed. Tools Appl. **77**(24), 31807–31833 (2018). https://doi.org/10.1007/s11042-018-6201-4
18. Al-Qershi, O.M., Khoo, B.E.: Enhanced block-based copy-move forgery detection using k-means clustering. Multidimens. Syst. Signal Process. **30**(4), 1671–1695 (2019). https://doi.org/10.1007/s11045-018-0624-y
19. Ambreen, I., Aatif, M., Jalil, Z., Iqbal, F., Marrington, A.: PViT: a hybrid model for deepfake face detection using patch vision transformers and deep learning. In: 2025 12th IFIP International Conference on New Technologies, Mobility and Security (NTMS), Paris, France, pp. 58–66. IEEE (2025). https://doi.org/10.1109/NTMS65597.2025.11076897
20. Amerini, I., Ballan, L., Caldelli, R., Del Bimbo, A., Del Tongo, L., Serra, G.: Copy-move forgery detection and localization by means of robust clustering with J-Linkage. Signal Process. Image Commun. **28**(6), 659–669 (2013). https://doi.org/10.1016/j.image.2013.03.006
21. Amerini, I., Ballan, L., Caldelli, R., Del Bimbo, A., Serra, G.: A SIFT-based forensic method for copy–move attack detection and transformation recovery. IEEE Trans. Inf. Forensics Secur. **6**(3), 1099–1110 (2011). https://doi.org/10.1109/TIFS.2011.2129512
22. Ardizzone, E., Bruno, A., Mazzola, G.: Copy-move forgery detection by matching triangles of keypoints. IEEE Trans. Inf. Forensics Secur. **10**(10), 2084–2094 (2015). https://doi.org/10.1109/TIFS.2015.2445742
23. Asghar, K., Habib, Z., Hussain, M.: Copy-Move and Splicing Image Forgery Detection and Localization Techniques: A Review. Taylor and Francis (2017). https://doi.org/10.1080/00450618.2016.1153711
24. Athitsos, V., Swain, M.J., Frankel, C.: Distinguishing photographs and graphics on the World Wide Web. In: 1997 Proceedings IEEE Workshop on Content-Based Access of Image and Video Libraries, St. Thomas, VI, USA, pp. 10–17. IEEE (1997). https://doi.org/10.1109/IVL.1997.629715
25. Autade, R., Gurajada, H.N.H.: Computer vision for financial fraud prevention using visual pattern analysis. In: 2025 International Conference on Engineering, Technology & Management (ICETM), Oakdale, NY, USA, pp. 1–7. IEEE (2025). https://doi.org/10.1109/ICETM63734.2025.11051811
26. Bappy, J.H., Roy-Chowdhury, A.K., Bunk, J., Nataraj, L., Manjunath, B.S.: Exploiting spatial structure for localizing manipulated image regions. In: 2017 IEEE International Conference on Computer Vision (ICCV), Venice, Italy, pp. 4980–4989. IEEE (2017). https://doi.org/10.1109/ICCV.2017.532
27. Bappy, J.H., Simons, C., Nataraj, L., Manjunath, B.S., Roy-Chowdhury, A.K.: Hybrid LSTM and encoder–decoder architecture for detection of image forgeries. IEEE Trans. Image Process. **28**(7), 3286–3300 (2019). https://doi.org/10.1109/TIP.2019.2895466

28. Baraheem, S.S., Nguyen, T.V.: AI vs. AI: can AI detect AI-generated images? J. Imaging. **9**(10), 199 (2023). https://doi.org/10.3390/jimaging9100199
29. Barnes, C., Shechtman, E., Finkelstein, A., Goldman, D.B.: PatchMatch: a randomized correspondence algorithm for structural image editing. ACM Trans. Graph. (2009). https://doi.org/10.1145/1531326.1531330
30. Barni, M., Phan, Q.-T., Tondi, B.: Copy move source-target disambiguation through multi-branch CNNs. IEEE Trans. Inf. Forensics Secur. **16**, 1825–1840 (2021). https://doi.org/10.1109/TIFS.2020.3045903
31. Baru, L.B., Boddeda, R., Patel, S.A., Gajapaka, S.M.: Wavelet-driven generalizable framework for deepfake face forgery detection. In: 2025 IEEE/CVF Winter Conference on Applications of Computer Vision Workshops (WACVW), Tucson, AZ, USA, pp. 1576–1584. IEEE (2025). https://doi.org/10.1109/WACVW65960.2025.00180
32. Bas, P., Filler, T., Pevný, T.: "Break our steganographic system": the ins and outs of organizing BOSS. In: Information Hiding. IH 2011. Lecture Notes in Computer Science, vol. 6958, pp. 59–70. Springer, Berlin, Heidelberg (2011). https://doi.org/10.1007/978-3-642-24178-9_5
33. Bay, H., Ess, A., Tuytelaars, T., Van Goo, L.: Speeded-up robust features (surf). Comput. Vis. Image Underst. **110**(3), 346–359 (2008). https://doi.org/10.1016/j.cviu.2007.09.014
34. Bayar, B., Stamm, M.C.: Constrained convolutional neural networks: a new approach towards general purpose image manipulation detection. IEEE Trans. Inf. Forensics Secur. **13**(11), 2691–2706 (2018). https://doi.org/10.1109/TIFS.2018.2825953
35. Bayram, S., Taha Sencar, H., Memon, N.: An efficient and robust method for detecting copy-move forgery. In: 2009 IEEE International Conference on Acoustics, Speech and Signal Processing, Taipei, Taiwan, pp. 1053–1056. IEEE (2009). https://doi.org/10.1109/ICASSP.2009.4959768
36. Bevinamarad, P., Unki, P., Nidagundi, P.: Copy-move forgery detection and localization framework for images using stationary wavelet transform and hybrid dilated adaptive VGG16 with optimization strategy. Int. J. Image Graph. Signal Process. **16**(1), 38–60 (2024). https://doi.org/10.5815/ijigsp.2024.01.04
37. Bhowal, A., Neogy, S., Naskar, R.: Deep learning-based forgery detection and localization for compressed images using a hybrid optimization model. Multimed. Syst. **30**, 128 (2024). https://doi.org/10.1007/s00530-024-01336-6
38. Bi, X., Liu, Y., Xiao, B., Li, W., Pun, C.-M., Wang, G., Gao, X.: D-Unet: a dual-encoder U-Net for image splicing forgery detection and localization. arXiv. (2020). https://doi.org/10.48550/arXiv.2012.01821
39. Bi, X., Pun, C.M.: Fast copy-move forgery detection using local bidirectional coherency error refinement. Pattern Recogn. **81**, 161–175 (2018). https://doi.org/10.1016/j.patcog.2018.03.028
40. Bi, X., Wei, Y., Xiao, B., Li, W.: RRU-Net: the ringed residual U-Net for image splicing forgery detection. In: 2019 IEEE/CVF Conference on Computer Vision and Pattern Recognition Workshops (CVPRW), Long Beach, CA, USA, pp. 30–39. IEEE (2019). https://doi.org/10.1109/CVPRW.2019.00010
41. Bilal, M., Habib, H.A., Mehmood, Z., Yousaf, R.M., Saba, T., Rehman, A.: A robust technique for copy-move forgery detection from small and extremely smooth tampered regions based on the DHE-SURF features and mDBSCAN clustering. Aust. J. Forensic Sci. **53**(4), 459–482 (2020). https://doi.org/10.1080/00450618.2020.1715479
42. Birajdar, G.K., Mankar, V.H.: Digital image forgery detection using passive techniques: a survey. Digit. Investig. **10**(3), 226–245 (2013). https://doi.org/10.1016/j.diin.2013.04.007
43. Birajdar, G.K., Mankar, V.H.: Computer graphic and photographic image classification using local image descriptors. Def. Sci. J. (2017). https://doi.org/10.14429/DSJ.67.10079
44. Bunk, J., et al.: Detection and localization of image forgeries using resampling features and deep learning. In: 2017 IEEE Conference on Computer Vision and Pattern Recognition Workshops (CVPRW), Honolulu, HI, USA, pp. 1881–1889. IEEE (2017). https://doi.org/10.1109/CVPRW.2017.235

45. Cao, G., Chen, Y., Zong, G., Chen, Y.: Detection of copy-move forgery in digital image using locality preserving projections. In: 2015 8th International Congress on Image and Signal Processing (CISP), Shenyang, China, pp. 599–603 (2015). https://doi.org/10.1109/CISP.2015.7407949
46. Cao, G., Zhao, Y., Ni, R., Li, X.: Contrast enhancement-based forensics in digital images. IEEE Trans. Inf. Forensics Secur. **9**(3), 515–525 (2014). https://doi.org/10.1109/TIFS.2014.2300937
47. de Carvalho, T.J., Riess, C., Angelopoulou, E., Pedrini, H., de Rezende Rocha, A.: Exposing digital image forgeries by illumination color classification. IEEE Trans. Inf. Forensics Secur. **8**(7), 1182–1194 (2013). https://doi.org/10.1109/TIFS.2013.2265677
48. Chauhan, D., Kasat, D., Jain, S., Thakare, V.: Survey on keypoint based copy-move forgery detection methods on image. In: Procedia Computer Science, pp. 206–212. Elsevier B.V (2016). https://doi.org/10.1016/j.procs.2016.05.213
49. Chen, Z., Li, H., Hu, J.: Multi-scale high-frequency vision transformer for face forgery detection. IEEE Trans. Biom. Behav. Identity Sci. https://doi.org/10.1109/TBIOM.2025.3543692
50. Chen, Y., Li, Z., Li, M., Ma, W.: Automatic classification of photographs and graphics. In: 2006 IEEE International Conference on Multimedia and Expo, Toronto, ON, Canada, pp. 973–976. IEEE (2006). https://doi.org/10.1109/ICME.2006.262695
51. Chen, D., Li, J., Wang, S., Li, S.: Identifying computer generated and digital camera images using fractional lower order moments. In: 2009 4th IEEE Conference on Industrial Electronics and Applications, Xi'an, pp. 230–235. IEEE (2009). https://doi.org/10.1109/ICIEA.2009.5138202
52. Chen, J., Liao, X., Qin, Z.: Identifying tampering operations in image operator chains based on decision fusion. Signal Process. Image Commun. **95**, 116287 (2021). https://doi.org/10.1016/j.image.2021.116287
53. Chen, J., Lo, M., Liao, H., Huang, T.: IPD-Net: detecting AI-generated images via inter-patch dependencies. Int. J. Adv. Comput. Sci. Appl. **15**(7) (2024). https://doi.org/10.14569/IJACSA.2024.01507133
54. Chen, C., McCloskey, S., Yu, J.: Image splicing detection via camera response function analysis. In: 2017 IEEE Conference on Computer Vision and Pattern Recognition (CVPR), Honolulu, HI, USA, pp. 1876–1885. IEEE (2017). https://doi.org/10.1109/CVPR.2017.203
55. Chen, B., Qi, X., Sun, X., Shi, Y.-Q.: Quaternion pseudo-Zernike moments combining both of RGB information and depth information for color image splicing detection. J. Vis. Commun. Image Represent. **49**, 283–290 (2017). https://doi.org/10.1016/j.jvcir.2017.08.011
56. Chen, B., Qi, X., Wang, Y., Zheng, Y., Shim, H.J., Shi, Y.-Q.: An improved splicing localization method by fully convolutional networks. IEEE Access. **6**, 69472–69480 (2018). https://doi.org/10.1109/ACCESS.2018.2880433
57. Chen, W., Shi, Y.Q., Xuan, G.: Identifying computer graphics using hsv color model and statistical moments of characteristic functions. In: 2007 IEEE International Conference on Multimedia and Expo, Beijing, China, pp. 1123–1126. IEEE (2007). https://doi.org/10.1109/ICME.2007.4284852
58. Chen, W., Shi, Y.Q., Xuan, G., Su, W.: Computer graphics identification using genetic algorithm. In: 2008 19th International Conference on Pattern Recognition, Tampa, FL, USA, pp. 1–4. IEEE (2008). https://doi.org/10.1109/ICPR.2008.4761552
59. Chen, H., Yang, X., Lyu, Y.: Copy-move forgery detection based on keypoint clustering and similar neighborhood search algorithm. IEEE Access. **8**, 36863–36875 (2020). https://doi.org/10.1109/ACCESS.2020.2974804
60. Chen, H., Zhao, C., Shi, Z., Zhu, F.: An image splicing localization algorithm based on SLIC and image features. In: Hong, R., Cheng, W.H., Yamasaki, T., Wang, M., Ngo, C.W. (eds.) Advances in Multimedia Information Processing – PCM 2018. PCM 2018 Lecture Notes in Computer Science, vol. 11166. Springer, Cham (2018). https://doi.org/10.1007/978-3-030-00764-5_56

61. Chen, Y., et al.: A semantically impactful image manipulation dataset: characterizing image manipulations using semantic significance. In: 2025 IEEE/CVF Winter Conference on Applications of Computer Vision (WACV), Tucson, AZ, USA, pp. 7659–7668. IEEE (2025). https://doi.org/10.1109/WACV61041.2025.00744
62. Chen, C., et al.: Moiré spectral augmentation and masked frequency modeling for document presentation attack detection. IEEE Trans. Dependable Secure Comput. https://doi.org/10.1109/TDSC.2025.3566310
63. Cheng, J., et al.: ED4: explicit data-level debiasing for deepfake detection. IEEE Trans. Image Process. **34**, 4618–4630 (2025). https://doi.org/10.1109/TIP.2025.3588323
64. Chollet, F.: Xception: deep learning with depthwise separable convolutions. In: 2017 IEEE Conference on Computer Vision and Pattern Recognition (CVPR), pp. 1800–1807. IEEE (2017). https://doi.org/10.1109/CVPR.2017.195
65. Christlein, V., Riess, C., Angelopoulou, E.: On rotation invariance in copy-move forgery detection. In: 2010 IEEE International Workshop on Information Forensics and Security, pp. 1–6. IEEE (2010). https://doi.org/10.1109/WIFS.2010.5711472
66. Christlein, V., Riess, C., Jordan, J., Riess, C., Angelopoulou, E.: An evaluation of popular copy-move forgery detection approaches. IEEE Trans. Inf. Forensics Secur. **7**(6), 1841–1854 (2012). https://doi.org/10.1109/TIFS.2012.2218597
67. Conotter, V., Cordin, L.: Detecting photographic and computer generated composites. In: Proc. SPIE 7870, Image Processing: Algorithms and Systems IX, 78700A. SPIE (2011). https://doi.org/10.1117/12.872565
68. Cozzolino, D., Poggi, G., Verdoliva, L.: Copy-move forgery detection based on PatchMatch. In: 2014 IEEE International Conference on Image Processing, ICIP 2014, pp. 5312–5316. IEEE (2014). https://doi.org/10.1109/ICIP.2014.7026075
69. Cozzolino, D., Poggi, G., Verdoliva, L.: Efficient dense-field copy-move forgery detection. IEEE Trans. Inf. Forensics Secur. **10**(11), 2284–2297 (2015). https://doi.org/10.1109/TIFS.2015.2455334
70. Cozzolino, D., Verdoliva, L.: Noiseprint: a CNN-based camera model fingerprint. IEEE Trans. Inf. Forensics Secur. **15**, 144–159 (2020). https://doi.org/10.1109/TIFS.2019.2916364
71. Cui, S., Nguyen, H.H., Le, T.-N., Lu, C.-S., Echizen, I.: LookupForensics: a large-scale multi-task dataset for multi-phase image-based fact verification. IEEE Access. https://doi.org/10.1109/ACCESS.2025.3584395
72. Cun, X., Pun, C.M.: Image splicing localization via semi-global network and fully connected conditional random fields. In: Leal-Taixé, L., Roth, S. (eds.) Computer Vision – ECCV 2018 Workshops. ECCV 2018 Lecture Notes in Computer Science, vol. 11130. Springer, Cham (2019). https://doi.org/10.1007/978-3-030-11012-3_22
73. Dang-Nguyen, D.-T., Boato, G., De Natale, F.G.B.: Discrimination between computer generated and natural human faces based on asymmetry information. In: 2012 Proceedings of the 20th European Signal Processing Conference (EUSIPCO), Bucharest, Romania, pp. 1234–1238. IEEE (2012)
74. Das, A.M., Aji, S.: A fast and efficient method for image splicing localization using BM3D noise estimation. In: Krishna, A., Srikantaiah, K., Naveena, C. (eds.) Integrated Intelligent Computing, Communication and Security. Studies in Computational Intelligence, vol. 771. Springer, Singapore (2019). https://doi.org/10.1007/978-981-10-8797-4_65
75. Davarzani, R., Yaghmaie, K., Mozaffari, S., Tapak, M.: Copy-move forgery detection using multiresolution local binary patterns. Forensic Sci. Int. **231**(1–3), 61–72 (2013). https://doi.org/10.1016/j.forsciint.2013.04.023
76. Dehnie, S., Sencar, T., Memon, N.: Digital image forensics for identifying computer generated and digital camera images. In: 2006 International conference on image processing, Atlanta, GA, USA, pp. 2313–2316. IEEE (2006). https://doi.org/10.1109/ICIP.2006.312849
77. Deng, K., Deng, K.: FSDNet: a decoupled forgery style network for image forgery detection. In: 2025 6th International Conference on Computer Vision, Image and Deep

Learning (CVIDL), Ningbo, China, pp. 458–462. IEEE (2025). https://doi.org/10.1109/CVIDL65390.2025.11085528
78. Devagiri, V.M., Cheddad, A.: Splicing forgery detection and the impact of image resolution. In: 2017 9th International Conference on Electronics, Computers and Artificial Intelligence (ECAI), Targoviste, Romania, pp. 1–6 (2017). https://doi.org/10.1109/ECAI.2017.8166431
79. Dhivya, S., Sangeetha, J., Sudhakar, B.: Copy-move forgery detection using SURF feature extraction and SVM supervised learning technique. Soft. Comput. **24**(19), 14429–14440 (2020). https://doi.org/10.1007/s00500-020-04795-x
80. Dimitri, G.M., Tondi, B., Barni, M.: Enhancing synthetic generated-images detection through post-hoc calibration. In: 2025 IEEE/CVF Winter Conference on Applications of Computer Vision Workshops (WACVW), Tucson, AZ, USA, pp. 729–736. IEEE (2025). https://doi.org/10.1109/WACVW65960.2025.00087
81. Diwan, A., Kumar, D., Mahadeva, R., Perera, H.C.S., Alawatugoda, J.: Unveiling copy-move forgeries: enhancing detection with superpoint keypoint architecture. IEEE Access. **11**, 86132–86148 (2023). https://doi.org/10.1109/ACCESS.2023.3304728
82. Diwan, A., Roy, A.K.: CNN-keypoint based two-stage hybrid approach for copy-move forgery detection. IEEE Access. **12**, 43809–43826 (2024). https://doi.org/10.1109/ACCESS.2024.3380460
83. Diwan, A., Sharma, R., Roy, A.K., Mitra, S.K.: Keypoint based comprehensive copy-move forgery detection. IET Image Process. **15**(6), 1298–1309 (2021). https://doi.org/10.1049/ipr2.12105
84. Dixit, R., Naskar, R.: Review, analysis and parameterisation of techniques for copy-move forgery detection in digital images. IET Image Process. **11**(9), 746–759 (2017). https://doi.org/10.1049/iet-ipr.2016.0322
85. Dixit, R., Naskar, R., Mishra, S.: Blur-invariant copy-move forgery detection technique with improved detection accuracy utilising SWT-SVD. IET Image Process. **11**(5), 301–309 (2017). https://doi.org/10.1049/iet-ipr.2016.0537
86. El Biach, F.Z., Iala, I., Laanaya, H., Minaoui, K.: Encoder-decoder based convolutional neural networks for image forgery detection. Multimed. Tools Appl. **81**(16), 22611–22628 (2022). https://doi.org/10.1007/s11042-020-10158-3
87. Elaskily, M.A., Alkinani, M.H., Sedik, A., Dessouky, M.M.: Deep learning based algorithm (ConvLSTM) for copy move forgery detection. J. Intell. Fuzzy Syst. **40**(3), 4385–4405 (2021). https://doi.org/10.3233/JIFS-201192
88. Elaskily, M.A., et al.: A novel deep learning framework for copy-move forgery detection in images. Multimed. Tools Appl. **79**(27–28), 19167–19192 (2020). https://doi.org/10.1007/s11042-020-08751-7
89. El-Latif, E.I.A., Taha, A., Zayed, H.H.: A passive approach for detecting image splicing using deep learning and Haar Wavelet Transform. Int. J. Comput. Netw. Inf. Secur. **11**, 28–35 (2019). https://doi.org/10.5815/ijcnis.2019.05.04
90. Elsharkawy, Z.F., Abdelwahab, S.A.S., Abd El-Samie, F.E., et al.: New and efficient blind detection algorithm for digital image forgery using homomorphic image processing. Multimed. Tools Appl. **78**, 21585–21611 (2019). https://doi.org/10.1007/s11042-019-7206-3
91. Emam, M., Han, Q., Li, Q., Zhang, H., Emam, M.: A robust detection algorithm for image Copy-Move forgery in smooth regions. In: 2017 International Conference on Circuits, System and Simulation (ICCSS), London, UK, pp. 119–123. IEEE (2017). https://doi.org/10.1109/CIRSYSSIM.2017.8023194
92. Emam, M., Han, Q., Niu, X.: PCET based copy-move forgery detection in images under geometric transforms. Multimed. Tools Appl. **75**(18), 11513–11527 (2016). https://doi.org/10.1007/s11042-015-2872-2
93. Emam, M., Han, Q., Zhang, H.: Two-stage keypoint detection scheme for region duplication forgery detection in digital images. J. Forensic Sci. **63**(1), 102–111 (2018). https://doi.org/10.1111/1556-4029.13456

94. Eykholt, K., Evtimov, I., Fernandes, E., Li, B., Rahmati, A., Xiao, C., Prakash, A., Kohno, T., Song, D.X.: Robust physical-world attacks on deep learning visual classification. In: 2018 IEEE/CVF Conference on Computer Vision and Pattern Recognition, pp. 1625–1634. IEEE (2018)
95. Fadl, S.M., Semary, N.A.: A proposed accelerated image copy-move forgery detection. In: 2014 IEEE Visual Communications and Image Processing Conference, Valletta, Malta, pp. 253–257. IEEE (2014). https://doi.org/10.1109/VCIP.2014.7051552
96. Fan, S., Wang, R., Zhang, Y., Guo, K.: Classifying computer generated graphics and natural images based on image contour information. J. Inf. Comput. Sci. **9**, 2877–2895 (2012)
97. Farid, H.: Detecting digital forgeries using bispectral analysis. AI Lab, Massachusetts Institute of Technology, Tech Rep AIM-1657 (1999)
98. Farid, H.: Image forgery detection: a survey. Signal Proc. Mag. IEEE. **26**(04), 16–25 (2009). https://doi.org/10.1109/MSP.2008.931079
99. Fontani, M., Bianchi, T., De Rosa, A., Piva, A., Barni, M.: A framework for decision fusion in image forensics based on Dempster–Shafer theory of evidence. IEEE Trans. Inf. Forensics Secur. **8**(4), 593–607 (2013). https://doi.org/10.1109/TIFS.2013.2248727
100. Fridrich, J., Chen, M., Goljan, M.: Imaging sensor noise as digital X-ray for revealing forgeries. In: Information Hiding. IH 2007. Lecture Notes in Computer Science, vol. 4567. Springer, Berlin, Heidelberg (2007). https://doi.org/10.1007/978-3-540-77370-2_23
101. Fridrich, J., Soukal, D., Lukáš, J.: Detection of copy-move forgery in digital images. Proc. Digit. Forensic Res. Workshop. **3**(2) (2003)
102. Fridrich, J., Soukal, D., Lukás, J.: Detection of copy move forgery in digital images. In: Proc. Digital Forensic Research Workshop (2003)
103. Gallagher, A.C., Chen, T.: Image authentication by detecting traces of demosaicing. In: 2008 IEEE Computer Society Conference on Computer Vision and Pattern Recognition Workshops, Anchorage, AK, USA, pp. 1–8. IEEE (2008). https://doi.org/10.1109/CVPRW.2008.4562984
104. Gan, Y., Yang, J.: An effective scheme for copy-move forgery detection using polar sine transform. In: Proceedings - 2019 2nd International Conference on Safety Produce Informatization, IICSPI 2019, pp. 337–341. IEEE (2019). https://doi.org/10.1109/IICSPI48186.2019.9096005
105. Gani, G., Qadir, F.: A robust copy-move forgery detection technique based on discrete cosine transform and cellular automata. J. Inf. Secur. Appl. **54** (2020). https://doi.org/10.1016/j.jisa.2020.102510
106. Gao, M., et al.: Exploring acoustic reverse nonlinearity against speech forgery in real-time voice applications. In: IEEE INFOCOM 2025 - IEEE Conference on Computer Communications, London, United Kingdom, pp. 1–10. IEEE (2025). https://doi.org/10.1109/INFOCOM55648.2025.11044526
107. Ghorbani, M., Firouzmand, M., Faraahi, A.: DWT-DCT (QCD) based copy-move image forgery detection. In: 2011 18th International Conference on Systems, Signals and Image Processing, Sarajevo, Bosnia and Herzegovina, pp. 1–4. IEEE (2011)
108. Goel, N., Kaur, S., Bala, R.: Dual branch convolutional neural network for copy move forgery detection. IET Image Process. **15**(3), 656–665 (2021). https://doi.org/10.1049/ipr2.12051
109. Guan, W., Wang, W., Peng, B., He, Z., Dong, J., Cheng, H.: Noise-informed diffusion-generated image detection with anomaly attention. IEEE Trans. Inf. Forensics Secur. **20**, 5256–5268 (2025). https://doi.org/10.1109/TIFS.2025.3573161
110. Guo, K., Wang, R.: A new method for detecting computer-generated images based on multiwavelets. J. Inf. Comput. Sci. **8**, 1449–1456 (2011)
111. GuorenYao, G.Y., Liu, X., Chen, L.: Two-stream manipulation trace network for face forgery detection. In: 2025 8th International Conference on Advanced Electronic Materials, Computers and Software Engineering (AEMCSE), Nanjing, China, pp. 703–707. IEEE (2025). https://doi.org/10.1109/AEMCSE65292.2025.11042665
112. Haider, K., Abbas, M.S., Muhammad, A., Qaiser, M.: Enhanced synthetic images classification with vision transformers and effective data augmentation techniques. In: 2024 26th

International Multi-Topic Conference (INMIC), Karachi, Pakistan, pp. 1–6. IEEE (2024). https://doi.org/10.1109/INMIC64792.2024.11004341
113. Han, J.G., Park, T.H., Moon, Y.H., et al.: Quantization-based Markov feature extraction method for image splicing detection. Mach. Vis. Appl. **29**, 543–552 (2018). https://doi.org/10.1007/s00138-018-0911-5
114. He, M.: Distinguish computer generated and digital images: a CNN solution. Concurr. Comput. Pract. Exp. **4788**, 1–10 (2018). https://doi.org/10.1002/cpe.4788
115. He, P., Jiang, X., Sun, T., Li, H.: Computer graphics identification combining convolutional and recurrent neural networks. IEEE Signal Process. Lett. **25**(9), 1369–1373 (2018). https://doi.org/10.1109/LSP.2018.2855566
116. Hilal, A., Hamzeh, T., Chantaf, S.: Copy-move forgery detection using principal component analysis and discrete cosine transform. In: 2017 Sensors Networks Smart and Emerging Technologies (SENSET), pp. 1–4. IEEE (2017). https://doi.org/10.1109/SENSET.2017.8125021
117. Holmes, O., Banks, M.S., Farid, H.: Assessing and improving the identification of computer-generated portraits. ACM Trans. Appl. Percept. **13**(2), Article 7 (2016). https://doi.org/10.1145/2871714
118. Horváth, H.J., Montserrat, D.M., Hao, H., Delp, E.J.: Manipulation detection in satellite images using deep belief networks. In: 2020 IEEE/CVF Conference on Computer Vision and Pattern Recognition Workshops (CVPRW), Seattle, WA, USA, pp. 2832–2840. IEEE (2020). https://doi.org/10.1109/CVPRW50498.2020.00340
119. Hosny, K.M., Hamza, H.M., Lashin, N.A.: Copy-move forgery detection of duplicated objects using accurate PCET moments and morphological operators. Imaging Sci. J. **66**(6), 330–345 (2018). https://doi.org/10.1080/13682199.2018.1461345
120. Hosny, K.M., Hamza, H.M., Lashin, N.A.: Copy-for-duplication forgery detection in colour images using QPCETMs and sub-image approach. IET Image Process. **13**(9), 1437–1446 (2019). https://doi.org/10.1049/iet-ipr.2018.5356
121. Hsu, Y., Chang, S.: Detecting image splicing using geometry invariants and camera characteristics consistency. In: 2006 IEEE International Conference on Multimedia and Expo, Toronto, ON, Canada, pp. 549–552. IEEE (2006). https://doi.org/10.1109/ICME.2006.262447
122. Huang, D.Y., Huang, C.N., Hu, W.C., Chou, C.H.: Robustness of copy-move forgery detection under high JPEG compression artifacts. Multimed. Tools Appl. **76**(1), 1509–1530 (2017). https://doi.org/10.1007/s11042-015-3152-x
123. Hussain, M., Muhammad, G., Saleh, S.Q., Mirza, A.M., Bebis, G.: Copy-move image forgery detection using multi-resolution Weber descriptors. In: 8th International Conference on Signal Image Technology and Internet Based Systems, SITIS 2012, pp. 395–401. IEEE (2012). https://doi.org/10.1109/SITIS.2012.64
124. Hussain, M., Saleh, S.Q., Aboalsamh, H., Muhammad, G., Bebis, G.: Comparison between WLD and LBP descriptors for non-intrusive image forgery detection. In: 2014 IEEE International Symposium on Innovations in Intelligent Systems and Applications (INISTA) Proceedings, Alberobello, Italy, pp. 197–204. IEEE (2014). https://doi.org/10.1109/INISTA.2014.6873618
125. Hussien, N.Y., Mahmoud, R.O., Zayed, H.H.: Deep learning on digital image splicing detection using CFA artifacts. Int. J. Sociotechnol. Knowl. Dev. **12**, 31–44 (2020). https://doi.org/10.4018/IJSKD.2020040102
126. Huynh, T.K., Huynh, K.V., Le-Tien, T., Nguyen, S.C.: A survey on image forgery detection techniques. In: The 2015 IEEE RIVF international conference on computing & communication technologies-research, innovation, and vision for future (RIVF), pp. 71–76. IEEE (2015). https://doi.org/10.1109/RIVF.2015.7049877
127. Huynh, K.-T., Ly, T.-N., Le-Tien, T.: ORB for detecting copy-move regions with scale and rotation in image forensics. In: Future data and security engineering. Big data, security and privacy, smart city and Industry 4.0 applications, pp. 358–372. Springer, Singapore (2020). https://doi.org/10.1007/978-981-33-4370-2_25

128. Ilhan, I., Karaköse, M.: MesoNet-ViT: Meso network and vision transformer for deepfake detection. In: 2025 29th International Conference on Information Technology (IT), Zabljak, Montenegro, pp. 1–4. IEEE (2025). https://doi.org/10.1109/IT64745.2025.10930249
129. İmamoğlu, M.B., Ulutaş, G., Ulutaş, M.: Detection of copy-move forgery using Krawtchouk moment. In: 2013 8th International Conference on Electrical and Electronics Engineering (ELECO), Bursa, Turkey, pp. 311–314. IEEE (2013). https://doi.org/10.1109/ELECO.2013.6713851
130. Isaac, M.M., Wilscy, M.: Copy-move forgery detection based on Harris Corner points and BRISK. In: Proceedings of the Third International Symposium on Women in Computing and Informatics (WCI '15), pp. 394–399. Association for Computing Machinery, New York, NY (2015). https://doi.org/10.1145/2791405.2791453
131. Itier, V., Strauss, O., Morel, L., Puech, W.: Color noise correlation-based splicing detection for image forensics. Multimed. Tools Appl. **80** (2021)
132. Jalab, H.A., Subramaniam, T., Ibrahim, R.W., Kahtan, H., Noor, N.F.M.: New texture descriptor based on modified fractional entropy for digital image splicing forgery detection. Entropy. **21**(4), 371 (2019). https://doi.org/10.3390/e21040371
133. Jin, Z., Zeng, Y., Liu, T., Liu, X., Leng, B., Zhou, X.: Protecting consumer electronics human-computer interactive verification security via anomaly-aware reconstruction-guided forgery localization. IEEE Trans. Consum. Electron. https://doi.org/10.1109/TCE.2025.3566783
134. Johnson, M.K., Farid, H.: Exposing digital forgeries by detecting inconsistencies in lighting. In: Proceedings of the ACM Multimedia and Security Workshop, pp. 1–10. ACM, New York, NY (2005). https://doi.org/10.1145/073170107317
135. Johnson, M.K., Farid, H.: Exposing digital forgeries through chromatic aberration. In: Proceedings of the ACM Multimedia and Security Workshop, Geneva, pp. 48–55. ACM (2006). https://doi.org/10.1145/1161366.1161376
136. Johnson, M.K., Farid, H.: Detecting photographic composites of people. In: Proceedings of the 6th International Workshop on Digital Watermarking, Guangzhou. Springer-Verlag, Berlin, Heidelberg (2007). https://doi.org/10.1007/978-3-540-92238-4_3
137. Johnson, M.K., Farid, H.: Exposing digital forgeries through specular highlights on the eye. In: Information Hiding. IH 2007. Lecture Notes in Computer Science, vol. 4567, pp. 311–325. Springer, Berlin, Heidelberg (2007). https://doi.org/10.1007/978-3-540-77370-2_21
138. Jwaid, M.F., Baraskar, T.N.: Detection of copy-move image forgery using local binary pattern with discrete wavelet transform and principle component analysis. In: 2017 International Conference on Computing, Communication, Control and Automation (ICCUBEA), pp. 1–6. IEEE (2017). https://doi.org/10.1109/ICCUBEA.2017.8463695
139. Kakar, P., Sudha, N.: Exposing postprocessed copy-paste forgeries through transform-invariant features. IEEE Trans. Inf. Forensics Secur. **7**(3), 1018–1028 (2012). https://doi.org/10.1109/TIFS.2012.2188390
140. Ketenci, S., Ulutas, G.: Copy-move forgery detection in images via 2D-Fourier transform. In: 2013 36th International Conference on Telecommunications and Signal Processing (TSP), pp. 813–816. IEEE (2013). https://doi.org/10.1109/TSP.2013.6614051
141. Khanna, N., Chiu, G.T.-C., Allebach, J.P., Delp, E.J.: Forensic techniques for classifying scanner, computer generated and digital camera images. In: 2008 IEEE International Conference on Acoustics, Speech and Signal Processing, Las Vegas, NV, USA, pp. 1653–1656. IEEE (2008). https://doi.org/10.1109/ICASSP.2008.4517944
142. Khayeat, A., Sun, X., Rosin, P.: Improved DSIFT descriptor based copy-rotate-move forgery detection. In: Image and Video Technology. Springer (2015). https://doi.org/10.1007/978-3-319-29451-3
143. Korus, P.: Digital image integrity–a survey of protection and verification techniques. Digit. Signal Process. **71**, 1–26 (2017). https://doi.org/10.1016/j.dsp.2017.08.009
144. Korus, P., Huang, J.: Multi-scale fusion for improved localization of malicious tampering in digital images. IEEE Trans. Image Process. **25**(3), 1312–1326 (2016). https://doi.org/10.1109/TIP.2016.2518870

145. Korus, P., Huang, J.: Evaluation of random field models in multi-modal unsupervised tampering localization. In: 2016 IEEE international workshop on information forensics and security (WIFS), pp. 1–6. IEEE (2016). https://doi.org/10.1109/WIFS.2016.7823898
146. Kumar, S., Desai, J.V., Mukherjee, S.: A fast keypoint based hybrid method for copy move forgery detection. Int. J. Comput. Digit. Syst. **4**(2), 91–99 (2015). https://doi.org/10.12785/IJCDS/040203
147. Kumar, A., Singh, K.U., Swarup, C., Singh, T., Raja, L., Kumar, A.: Detection of copy-move forgery using Euclidean distance and texture features. Traitement du Signal. **39**(3), 781–788 (2022). https://doi.org/10.18280/ts.390302
148. Kumari, R., Garg, H.: An image copy-move forgery detection based on SURF and Fourier-Mellin transforms. In: 2023 International Conference on Artificial Intelligence and Smart Communication (AISC), pp. 515–519. IEEE (2023). https://doi.org/10.1109/AISC56616.2023.10085429
149. Kuznetsov, O., Frontoni, E., Romeo, L., Rosati, R.: Enhancing copy-move forgery detection through a novel CNN architecture and comprehensive dataset analysis. Multimed. Tools Appl. **83**(21), 59783–59817 (2024). https://doi.org/10.1007/s11042-023-17964-5
150. Lee, S.I., Park, J.Y., Eom, I.K.: CNN-based copy-move forgery detection using rotation-invariant wavelet feature. IEEE Access. **10**, 106217–106229 (2022). https://doi.org/10.1109/ACCESS.2022.3212069
151. Li, Y., He, Y., Chen, C., Dong, L., Li, B., Zhou, J., Li, X.: Image copy-move forgery detection via deep PatchMatch and pairwise ranking learning. IEEE Trans. Image Proc. **34**, 425–440 (2025). https://doi.org/10.1109/TIP.2024.3482191
152. Li, L., Li, S., Wang, J.: s. In: 2012 World Congress on Information and Communication Technologies, pp. 1061–1065. IEEE (2012). https://doi.org/10.1109/WICT.2012.6409232
153. Li, J., Li, X., Yang, B., Sun, X.: Segmentation-based image copy-move forgery detection scheme. IEEE Trans. Inf. Forensics Secur. **10**(3), 507–518 (2015). https://doi.org/10.1109/TIFS.2014.2381872
154. Li, L., Li, S., Zhu, H., Chu, S.-C., Roddick, J.F., Pan, J.-S.: An efficient scheme for detecting copy-move forged images by local binary patterns. J. Inf. Hiding Multimed. Signal Process. **4**(1), 46–56 (2013)
155. Li, L., Li, S., Zhu, H., Wu, X.: Detecting copy-move forgery under affine transforms for image forensics. Comput. Electr. Eng. **40**(6), 1951–1962 (2014). https://doi.org/10.1016/j.compeleceng.2013.11.034
156. Li, C., Ma, Q., Xiao, L., Li, M., Zhang, A.: Image splicing detection based on Markov features in QDCT domain. Neurocomputing. **228**, 29–36 (2017). https://doi.org/10.1016/j.neucom.2016.04.068
157. Li, J., Wang, K.: Detecting computer-generated images by using only real images. In: Proc. SPIE 13517, Seventeenth International Conference on Machine Vision (ICMV 2024), 135170Q (2025). https://doi.org/10.1117/12.3055092
158. Li, S.H., Wu, H., Wang, B.: A solution to ACMMM 2024 on artificial intelligence generated image detection. In: Proceedings of the 32nd ACM International Conference on Multimedia (MM '24), pp. 11475–11477. Association for Computing Machinery, New York, NY (2024). https://doi.org/10.1145/3664647.3689003
159. Li, Z., Ye, J., Shi, Y.Q.: Distinguishing computer graphics from photographic images using local binary patterns. In: Shi, Y.Q., Kim, H.J., Pérez-González, F. (eds.) The International Workshop on Digital Forensics and Watermarking 2012. IWDW 2012. Lecture Notes in Computer Science, vol. 7809. Springer, Berlin, Heidelberg (2013). https://doi.org/10.1007/978-3-642-40099-5_19
160. Li, W., Yu, N.: Rotation robust detection of copy-move forgery. In: 2010 IEEE International Conference on Image Processing, pp. 2113–2116. IEEE (2010). https://doi.org/10.1109/ICIP.2010.5652519
161. Li, W., Zhang, T., Zheng, E., Ping, X.: Identifying photorealistic computer graphics using second-order difference statistics. In: 2010 Seventh International Conference on Fuzzy

Systems and Knowledge Discovery, Yantai, China, pp. 2316–2319. IEEE (2010). https://doi.org/10.1109/FSKD.2010.5569821
162. Liang, E., Zhang, K., Hua, Z., Jia, X.: Dual-branch noise-guided network for image splicing forgery detection. IEEE Signal Process. Lett. https://doi.org/10.1109/LSP.2025.3589929
163. Liu, M., Di, X., Liao, M.: Image inpainting detection via dual guidance of uncertainty and precise boundary information. IEEE Trans. Circuits Syst. Video Technol. https://doi.org/10.1109/TCSVT.2025.3570481
164. Peng, F., Shi, J., Long, M.: Identifying photographic images and photorealistic computer graphics using multifractal spectrum features of PRNU. In: 2014 IEEE International Conference on Multimedia and Expo (ICME), Chengdu, China, pp. 1–6. IEEE (2014). https://doi.org/10.1109/ICME.2014.6890296
165. Peng, F., Zhou, D.-l., Long, M., Sun, X.-m.: Discrimination of natural images and computer generated graphics based on multi-fractal and regression analysis. AEU-Int. J. Electron. C. **71**, 72–81 (2017). https://doi.org/10.1016/j.aeue.2016.11.009
166. Peng, F., Zhou, D.-l.: Discriminating natural images and computer generated graphics based on the impact of CFA interpolation on the correlation of PRNU. Digit. Investig. **11**(2), 111–119 (2014). https://doi.org/10.1016/j.diin.2014.04.002
167. Nguyen, H.H., Ngoc-Dung Tieu, T., Nguyen-Son, H.-Q., Nozick, V., Yamagishi, J., Echizen, I.: Modular convolutional neural network for discriminating between computer-generated images and photographic images. In: Proceedings of the 13th International Conference on Availability, Reliability and Security (ARES '18), Article 1, pp. 1–10. Association for Computing Machinery, New York, NY (2018). https://doi.org/10.1145/3230833.3230863
168. Yu, I.-J., Kim, D.-G., Park, J.-S., Hou, J.-U., Choi, S., Lee, H.-K.: Identifying photorealistic computer graphics using convolutional neural networks. In: 2017 IEEE International Conference on Image Processing (ICIP), Beijing, China, pp. 4093–4097. IEEE (2017). https://doi.org/10.1109/ICIP.2017.8297052
169. Ng, T., Chang, S.: Classifying photographic and photorealistic computer graphic images using natural image statistics. ADVENT Technical Report #220-2006-6, Columbia University, New York, NY (2004)
170. Ng, T.T., Chang, S.F.: Discrimination of computer synthesized or recaptured images from real images. In: Sencar, H., Memon, N. (eds.) Digital Image Forensics. Springer, New York, NY (2013). https://doi.org/10.1007/978-1-4614-0757-7_10
171. Pan, F., Chen, J., Huang, J.: Discriminating between photorealistic computer graphics and natural images using fractal geometry. Sci. China Ser. F-Inf. Sci. **52**, 329–337 (2009). https://doi.org/10.1007/s11432-009-0053-5
172. Pan, F., Huang, J.: Discriminating computer graphics images and natural images using hidden Markov tree model. In: Kim, H.J., Shi, Y.Q., Barni, M. (eds.) Digital Watermarking. IWDW 2010 Lecture Notes in Computer Science, vol. 6526. Springer, Berlin, Heidelberg (2011). https://doi.org/10.1007/978-3-642-18405-5_3
173. Peng, F., Liu, J., Long, M.: Identification of natural images and computer generated graphics based on hybrid features. Int. J. Digit. Crime Forensics. **4**(1), 1–16 (2012). https://doi.org/10.4018/jdcf.2012010101
174. Ng, T.-T., Chang, S.-F.: An online system for classifying computer graphics images from natural photographs. In: Proc. SPIE 6072, Security, Steganography, and Watermarking of Multimedia Contents VIII, 607211. SPIE (2006). https://doi.org/10.1117/12.650162
175. Ng, T.-T., Chang, S.-F., Hsu, J., Xie, L., Tsui, M.-P.: Physics-motivated features for distinguishing photographic images and computer graphics. In: Proceedings of the 13th annual ACM international conference on Multimedia (MULTIMEDIA '05), pp. 239–248. Association for Computing Machinery, New York, NY (2005). https://doi.org/10.1145/1101149.1101192
176. Xu, B., Wang, J., Liu, G., et al.: Photorealistic computer graphics forensics based on leading digit law. J. Electron. (China). **28**, 95–100 (2011). https://doi.org/10.1007/s11767-011-0474-3

177. Wang, Y., Moulin, P.: On discrimination between photorealistic and photographic images. In: 2006 IEEE International Conference on Acoustics Speech and Signal Processing Proceedings, Toulouse. IEEE (2006). https://doi.org/10.1109/ICASSP.2006.1660304
178. Rahmouni, N., Nozick, V., Yamagishi, J., Echizen, I.: Distinguishing computer graphics from natural images using convolution neural networks. In: 2017 IEEE Workshop on Information Forensics and Security (WIFS), Rennes, France, pp. 1–6 (2017). https://doi.org/10.1109/WIFS.2017.8267647
179. de Rezende, E.R.S., Ruppert, G.C.S., Theóphilo, A., Tokuda, E.K., Carvalho, T.: Exposing computer generated images by using deep convolutional neural networks. Signal Process. Image Commun. **66**, 113–126 (2018). https://doi.org/10.1016/j.image.2018.04.006
180. Sankar, G., Zhao, V., Yang, Y.-H.: Feature based classification of computer graphics and real images. In: 2009 IEEE International Conference on Acoustics, Speech and Signal Processing, Taipei, Taiwan, pp. 1513–1516. IEEE (2009). https://doi.org/10.1109/ICASSP.2009.4959883
181. Sutthiwan, P., Cai, X., Shi, Y.Q., Zhang, H.: Computer graphics classification based on Markov process model and boosting feature selection technique. In: 2009 16th IEEE International Conference on Image Processing (ICIP), Cairo, Egypt, pp. 2913–2916. IEEE (2009). https://doi.org/10.1109/ICIP.2009.5413344
182. Sutthiwan, P., Ye, J., Shi, Y.Q.: An enhanced statistical approach to identifying photorealistic images. In: Ho, A.T.S., Shi, Y.Q., Kim, H.J., Barni, M. (eds.) Digital Watermarking. IWDW 2009 Lecture Notes in Computer Science, vol. 5703. Springer, Berlin, Heidelberg (2009). https://doi.org/10.1007/978-3-642-03688-0_28
183. Tan, D.Q., Shen, X.J., Qin, J., et al.: Detecting computer generated images based on local ternary count. Pattern Recognit. Image Anal. **26**, 720–725 (2016). https://doi.org/10.1134/S1054661816040167
184. Tokuda, E., Pedrini, H., Rocha, A.: Computer generated images vs. digital photographs: a synergetic feature and classifier combination approach. J. Vis. Commun. Image Represent. **24**(8), 1276–1292 (2013). https://doi.org/10.1016/j.jvcir.2013.08.009
185. Wang, J., Li, T., Luo, X., Shi, Y.-Q., Jha, S.K.: Identifying computer generated images based on quaternion central moments in color quaternion wavelet domain. IEEE Trans. Circuits Syst. Video Technol. **29**(9), 2775–2785 (2019). https://doi.org/10.1109/TCSVT.2018.2867786
186. Wang, J., Li, T., Shi, Y.Q., et al.: Forensics feature analysis in quaternion wavelet domain for distinguishing photographic images and computer graphics. Multimed. Tools Appl. **76**, 23721–23737 (2017). https://doi.org/10.1007/s11042-016-4153-0
187. Wang, X., Liu, Y., Bingchao, X., Li, L., Xue, J.: A statistical feature based approach to distinguish PRCG from photographs. Comput. Vis. Image Underst. **128**, 84–93 (2014). https://doi.org/10.1016/j.cviu.2014.07.007
188. Wu, J., Kamath, M.V., Poehlman, S.: Detecting differences between photographs and computer generated images. In: Proceedings of the 24th IASTED International Conference on Signal Processing, Pattern Recognition, and Applications (SPPRA'06), pp. 268–273. ACTA Press (2006)
189. Wu, R., Li, X., Yang, B.: Identifying computer generated graphics via histogram features. In: 2011 18th IEEE International Conference on Image Processing, Brussels, Belgium, pp. 1933–1936. IEEE (2011). https://doi.org/10.1109/ICIP.2011.6115849
190. Zhang, R.S., Quan, W.Z., Fan, L.B., et al.: Distinguishing computer-generated images from natural images using channel and pixel correlation. J. Comput. Sci. Technol. **35**, 592–602 (2020). https://doi.org/10.1007/s11390-020-0216-9
191. Yao, Y., Hu, W., Zhang, W., Wu, T., Shi, Y.-Q.: Distinguishing computer-generated graphics from natural images based on sensor pattern noise and deep learning. Sensors. **18**(4), 1296 (2018). https://doi.org/10.3390/s18041296
192. Zhang, R., Wang, R.D., Ng, T.T.: Distinguishing photographic images and photorealistic computer graphics using visual vocabulary on local image edges. In: Shi, Y.Q., Kim, H.J., Perez-Gonzalez, F. (eds.) Digital Forensics and Watermarking. IWDW 2011 Lecture

Notes in Computer Science, vol. 7128. Springer, Berlin, Heidelberg (2012). https://doi.org/10.1007/978-3-642-32205-1_24
193. Zhang, R., Wang, R.: Distinguishing photorealistic computer graphics from natural images by imaging features and visual features. In: 2011 International Conference on Electronics, Communications and Control (ICECC), Ningbo, China, pp. 226–229. IEEE (2011). https://doi.org/10.1109/ICECC.2011.6067631
194. Zhu, M., Chen, H., Yan, Q., Huang, X., Lin, G., Li, W., Zhijun, T., Hu, H., Hu, J., Wang, Y.: GenImage: a million-scale benchmark for detecting AI-generated image. In: Proceedings of the 37th International Conference on Neural Information Processing Systems (NIPS '23), Article 3398, pp. 77771–77782. Curran Associates, Red Hook, NY (2023)
195. Yao, Y., Zhang, Z., Ni, X., Shen, Z., Chen, L., Xu, D.: CGNet: detecting computer-generated images based on transfer learning with attention module. Signal Process. Image Commun. **105**, 116692 (2022). https://doi.org/10.1016/j.image.2022.116692
196. Lokner Lađević, A., Kramberger, T., Kramberger, R., Vlahek, D.: Detection of AI-generated synthetic images with a lightweight CNN. AI. **5**(3), 1575–1593 (2024). https://doi.org/10.3390/ai5030076
197. Muthaiah, U., Divya, A., Swarnalaxmi, T.N., Vidhyasagar, B.S.: A comparative review of AI-generated vs real images and classification techniques. In: 2024 4th International Conference on Ubiquitous Computing and Intelligent Information Systems (ICUIS), Gobichettipalayam, India, pp. 141–147. IEEE (2024). https://doi.org/10.1109/ICUIS64676.2024.10866220
198. Truong, P.-H., Nguyen, T.-D., Truong, X.-H., Nguyen, N.-H., Pham, D.-T.: Employing a CNN detector to identify AI-generated images and against attacks on AI systems. In: 2024 1st International Conference On Cryptography and Information Security (VCRIS), Hanoi, Vietnam, pp. 1–6. IEEE (2024). https://doi.org/10.1109/VCRIS63677.2024.10813439
199. Weir, S., Khan, M.S., Moradpoor, N., Ahmad, J.: Enhancing AI-generated image detection with a novel approach and comparative analysis. In: 2024 17th International Conference on Security of Information and Networks (SIN), Sydney, Australia, pp. 1–7. IEEE (2024). https://doi.org/10.1109/SIN63213.2024.10871900
200. Meena, K.B., Tyagi, V.: Methods to distinguish photorealistic computer generated images from photographic images: a review. In: Singh, M., Gupta, P., Tyagi, V., Flusser, J., Ören, T., Kashyap, R. (eds.) Advances in Computing and Data Sciences. ICACDS 2019. Communications in Computer and Information Science, vol. 1045. Springer, Singapore (2019). https://doi.org/10.1007/978-981-13-9939-8_7
201. Meena, K.B., Tyagi, V.: A novel method to distinguish photorealistic computer generated images from photographic images. In: 2019 Fifth International Conference on Image Information Processing (ICIIP), Shimla, India, pp. 385–390. IEEE (2019). https://doi.org/10.1109/ICIIP47207.2019.8985711
202. Meena, K.B., Tyagi, V.: A deep learning based method to discriminate between photorealistic computer generated images and photographic images. In: Singh, M., Gupta, P., Tyagi, V., Flusser, J., Ören, T., Valentino, G. (eds.) Advances in Computing and Data Sciences. ICACDS 2020. Communications in Computer and Information Science, vol. 1244. Springer, Singapore (2020). https://doi.org/10.1007/978-981-15-6634-9_20
203. Meena, K.B., Tyagi, V.: Distinguishing computer-generated images from photographic images using two-stream convolutional neural network. Appl. Soft Comput. **100**, 107025 (2021). https://doi.org/10.1016/j.asoc.2020.107025. [Elaskily, M.A., Dessouky, M.M., Faragallah, O.S., Sedik, A.: A survey on traditional and deep learning copy move forgery detection (CMFD) techniques. Multimed. Tools Appl. **82**(22), 34409–34435 (2023). https://doi.org/10.1007/s11042-023-14424-y]
204. Mahdian, B., Saic, S.: Detection of copy-move forgery using a method based on blur moment invariants. Forensic Sci. Int. **171**(2–3), 180–189 (2007). https://doi.org/10.1016/j.forsciint.2006.11.002

205. Ustubioglu, B., Ulutas, G., Ulutas, M., Nabiyev, V.V.: A new copy move forgery detection technique with automatic threshold determination. AEU-Int. J. Electron. C. **70**(8), 1076–1087 (2016). https://doi.org/10.1016/j.aeue.2016.05.005
206. Pun, C.M., Chung, J.L.: A two-stage localization for copy-move forgery detection. Inf. Sci. (N Y). **463–464**, 33–55 (2018). https://doi.org/10.1016/j.ins.2018.06.040
207. Meena, K.B., Tyagi, V.: A copy-move image forgery detection technique based on Gaussian-Hermite moments. Multimed. Tools Appl. **78**(23), 33505–33526 (2019). https://doi.org/10.1007/s11042-019-08082-2
208. Meena, K.B., Tyagi, V.: A hybrid copy-move image forgery detection technique based on Fourier-Mellin and scale invariant feature transforms. Multimed. Tools Appl. **79**, 8197–8212 (2020). https://doi.org/10.1007/s11042-019-08343-0
209. Warif, N.B.A., et al.: Copy-move forgery detection: survey, challenges and future directions. J. Netw. Comput. Appl. **75**, 259–278 (2016). https://doi.org/10.1016/J.JNCA.2016.09.008
210. Warif, N.B.A., Idris, M.Y.I., Wahab, A.W.A., Ismail, N.S.N., Salleh, R.: A comprehensive evaluation procedure for copy-move forgery detection methods: results from a systematic review. Multimed. Tools Appl. **81**(11), 15171–15203 (2022). https://doi.org/10.1007/s11042-022-12010-2
211. Pham, N.T., Park, C.S.: Toward deep-learning-based methods in image forgery detection: a survey. IEEE Access. **11**, 11224–11237 (2023). https://doi.org/10.1109/ACCESS.2023.3241837
212. Teerakanok, S., Uehara, T.: Copy-move forgery detection: a state-of-the-art technical review and analysis. IEEE Access. **7**, 40550–40568 (2019). https://doi.org/10.1109/ACCESS.2019.2907316
213. Mahmood, T., et al.: A survey on block based copy move image forgery detection techniques. In: 2015 International Conference on Emerging Technologies (ICET), pp. 1–6. IEEE (2015). https://doi.org/10.1109/ICET.2015.7389169
214. Wang, Y., Tian, L., Li, C.: LBP-SVD based copy move forgery detection algorithm. In: Proceedings - 2017 IEEE International Symposium on Multimedia, ISM 2017, pp. 553–556. Institute of Electrical and Electronics Engineers Inc. (2017). https://doi.org/10.1109/ISM.2017.108
215. Warbhe, A.D., Dharaskar, R.V., Thakare, V.M.: A scaling robust copy-paste tampering detection for digital image forensics. In: Procedia Computer Science, pp. 458–465. Elsevier B.V (2016). https://doi.org/10.1016/j.procs.2016.03.059
216. Moussa, A.M.: A fast and accurate algorithm for copy-move forgery detection. In: 2015 Tenth International Conference on Computer Engineering & Systems (ICCES), Cairo, Egypt, pp. 281–285. IEEE (2015). https://doi.org/10.1109/ICCES.2015.7393060
217. Parveen, A., Khan, Z.H., Ahmad, S.N.: Block-based copy–move image forgery detection using DCT. Iran J. Comput. Sci. **2**(2), 89–99 (2019). https://doi.org/10.1007/s42044-019-00029-y
218. Malviya, A.V., Ladhake, S.A.: Pixel based image forensic technique for copy-move forgery detection using auto color correlogram. In: Procedia Computer Science, pp. 383–390. Elsevier B.V (2016). https://doi.org/10.1016/j.procs.2016.03.050
219. Soni, B., Das, P.K., Thounaojam, D.M.: CMFD: A detailed review of block based and key feature based techniques in image copy move forgery detection. IET Image Process. **12**, 167–178 (2018). https://doi.org/10.1049/iet-ipr.2017.0441
220. Park, C.S., Kim, C., Lee, J., Kwon, G.R.: Rotation and scale invariant upsampled log-polar Fourier descriptor for copy-move forgery detection. Multimed. Tools Appl. **75**(23), 16577–16595 (2016). https://doi.org/10.1007/s11042-016-3575-z
221. Tralic, D., Grgic, S., Sun, X., Rosin, P.L.: Combining cellular automata and local binary patterns for copy-move forgery detection. Multimed. Tools Appl. **75**(24), 16881–16903 (2016). https://doi.org/10.1007/s11042-015-2961-2
222. Mohebbian, E., Hariri, M.: Increase the efficiency of DCT method for detection of copy-move forgery in complex and smooth images. In: 2015 2nd International Conference on Knowledge-Based Engineering and Innovation (KBEI), Tehran, Iran, pp. 436–440. IEEE (2015). https://doi.org/10.1109/KBEI.2015.7436084

223. Liu, G., Wang, J., Lian, S., Wang, Z.: A passive image authentication scheme for detecting region-duplication forgery with rotation. J. Netw. Comput. Appl. **34**(5), 1557–1565 (2011). https://doi.org/10.1016/j.jnca.2010.09.001
224. Muhammad, G., Hussain, M., Bebis, G.: Passive copy move image forgery detection using undecimated dyadic wavelet transform. Digit. Investig. **9**(1), 49–57 (2012). https://doi.org/10.1016/j.diin.2012.04.004
225. Muzaffer, G., Ulutas, G.: A fast and effective digital image copy move forgery detection with binarized SIFT. In: 2017 40th International Conference on Telecommunications and Signal Processing (TSP), Barcelona, Spain, pp. 595–598. IEEE (2017). https://doi.org/10.1109/TSP.2017.8076056
226. Muzaffer, G., Ulutas, G., Ustubioglu, B.: Copy move forgery detection with Quadtree decomposition segmentation. In: 2020 43rd International Conference on Telecommunications and Signal Processing, TSP 2020, pp. 208–211. IEEE (2020). https://doi.org/10.1109/TSP49548.2020.9163516
227. Wang, H., Wang, H.X., Sun, X.M., Qian, Q.: A passive authentication scheme for copy-move forgery based on package clustering algorithm. Multimed. Tools Appl. **76**(10), 12627–12644 (2017). https://doi.org/10.1007/s11042-016-3687-5
228. Zhong, J.L., Pun, C.M.: An end-to-end dense-InceptionNet for image copy-move forgery detection. IEEE Trans. Inf. Forensics Secur. **15**, 2134–2146 (2020). https://doi.org/10.1109/TIFS.2019.2957693
229. Ouyang, J., Liu, Y., Liao, M.: Robust copy-move forgery detection method using pyramid model and Zernike moments. Multimed. Tools Appl. **78**(8), 10207–10225 (2019). https://doi.org/10.1007/s11042-018-6605-1
230. Zhao, J., Guo, J.: Passive forensics for copy-move image forgery using a method based on DCT and SVD. Forensic Sci. Int. **233**(1), 158–166 (2013). https://doi.org/10.1016/j.forsciint.2013.09.013
231. Zheng, J., Liu, Y., Ren, J., Zhu, T., Yan, Y., Yang, H.: Fusion of block and keypoints based approaches for effective copy-move image forgery detection. Multidimens. Syst. Signal Process. **27**(4), 989–1005 (2016). https://doi.org/10.1007/s11045-016-0416-1
232. Zhong, J., Gan, Y.: Detection of copy–move forgery using discrete analytical Fourier–Mellin transform. Nonlinear Dyn. **84**(1), 189–202 (2016). https://doi.org/10.1007/s11071-015-2374-9
233. Zhong, J., Gan, Y., Young, J., Lin, P.: Copy move forgery image detection via discrete radon and polar complex exponential transform-based moment invariant features. Int. J. Pattern Recognit. Artif. Intell. **31**(2) (2017). https://doi.org/10.1142/S0218001417540052
234. Zhong, J., Gan, Y., Young, J., Huang, L., Lin, P.: A new block-based method for copy move forgery detection under image geometric transforms. Multimed. Tools Appl. **76**(13), 14887–14903 (2017). https://doi.org/10.1007/s11042-016-4201-9
235. Qiao, M., Sung, A.H., Liu, Q., Ribeiro, B.M.: A novel approach for detection of copy-move forgery. In: Proceedings of Fifth International Conference on Advanced Engineering Computing and Applications in Sciences (ADVCOMP). IARIA (2011)
236. Verma, M., Singh, D.: Survey on image copy-move forgery detection. Multimed. Tools Appl. **83**(8), 23761–23797 (2024). https://doi.org/10.1007/s11042-023-16455-x
237. Zandi, M., Mahmoudi-Aznaveh, A., Talebpour, A.: Iterative copy-move forgery detection based on a new interest point detector. IEEE Trans. Inf. Forensics Secur. **11**(11), 2499–2512 (2016). https://doi.org/10.1109/TIFS.2016.2585118
238. Warif, N.B.A., Idris, M.Y.I., Wahab, A.W.A., Salleh, R., Ismail, A.: CMF-iteMS: an automatic threshold selection for detection of copy-move forgery. Forensic Sci. Int. **295**, 83–99 (2019). https://doi.org/10.1016/j.forsciint.2018.12.004
239. Warif, N.B.A., Wahab, A.W.A., Idris, M.Y.I., Salleh, R., Othman, F.: SIFT-symmetry: a robust detection method for copy-move forgery with reflection attack. J. Vis. Commun. Image Represent. **46**, 219–232 (2017). https://doi.org/10.1016/j.jvcir.2017.04.004

240. Zheng, N., Wang, Y., Xu, M.: A LBP-based method for detecting copy-move forgery with rotation. In: Lecture Notes in Electrical Engineering, pp. 261–267 (2013). https://doi.org/10.1007/978-94-007-6738-6_33
241. Yang, P., Yang, G., Zhang, D.: Rotation invariant local binary pattern for blind detection of copy-move forgery with affine transform. In: Cloud Computing and Security: Second International Conference, ICCCS 2016. Lecture Notes in Computer Science, vol. 10040. Springer, Cham (2016). https://doi.org/10.1007/978-3-319-48674-1
242. Pandey, R.C., Agrawal, R., Singh, S.K., Shukla, K.K.: Passive copy move forgery detection using SURF, HOG and SIFT features. In: Advances in Intelligent Systems and Computing, pp. 659–666. Springer Verlag (2014). https://doi.org/10.1007/978-3-319-11933-5_74
243. Ryu, S.J., Lee, M.J., Lee, H.K.: Detection of copy-rotate-move forgery using Zernike moments. In: Böhme, R., Fong, P.W.L., Safavi-Naini, R. (eds.) Information Hiding. IH 2010. Lecture Notes in Computer Science, vol. 6387. Springer, Berlin, Heidelberg (2010). https://doi.org/10.1007/978-3-642-16435-4_5
244. Thajeel, S.A., Mahmood, A.S., Humood, W.R., Sulong, G.: Detection copy-move forgery in image via quaternion polar harmonic transforms. KSII Trans. Internet Inf. Syst. **13**(8), 4005–4025 (2019). https://doi.org/10.3837/tiis.2019.08.010
245. Sharma, S., Ghanekar, U.: A rotationally invariant texture descriptor to detect copy move forgery in medical images. In: Proceedings - 2015 IEEE International Conference on Computational Intelligence and Communication Technology, CICT 2015, pp. 795–798. IEEE (2015). https://doi.org/10.1109/CICT.2015.88
246. Mahmood, T., Irtaza, A., Mehmood, Z., Tariq Mahmood, M.: Copy–move forgery detection through stationary wavelets and local binary pattern variance for forensic analysis in digital images. Forensic Sci. Int. **279**, 8–21 (2017). https://doi.org/10.1016/j.forsciint.2017.07.037
247. Mahmood, T., Mehmood, Z., Shah, M., Saba, T.: A robust technique for copy-move forgery detection and localization in digital images via stationary wavelet and discrete cosine transform. J. Vis. Commun. Image Represent. **53**, 202–214 (2018). https://doi.org/10.1016/j.jvcir.2018.03.015
248. Wang, X., Liu, Y., Xu, H., et al.: Robust copy–move forgery detection using quaternion exponent moments. Pattern. Anal. Applic. **21**, 451–467 (2018). https://doi.org/10.1007/s10044-016-0588-1
249. Wang, X.Y., Li, S., Liu, Y.N., Niu, Y., Yang, H.Y., Zhou, Z.: A new keypoint-based copy-move forgery detection for small smooth regions. Multimed. Tools Appl. **76**(22), 23353–23382 (2017). https://doi.org/10.1007/s11042-016-4140-5
250. Liu, Y., Xia, C., Zhu, X., Xu, S.: Two-stage copy-move forgery detection with self deep matching and proposal SuperGlue. IEEE Trans. Image Process. **31**, 541–555 (2022). https://doi.org/10.1109/TIP.2021.3132828
251. Zhang, Y., et al.: CNN-transformer based generative adversarial network for copy-move source/target distinguishment. IEEE Trans. Circuits Syst. Video Technol. **33**(5), 2019–2032 (2023). https://doi.org/10.1109/TCSVT.2022.3220630
252. Zhang, Z., Wang, C., Zhou, X.: A survey on passive image copy-move forgery detection. J. Inf. Process. Syst. **14**(1), 6–31 (2018). https://doi.org/10.3745/JIPS.02.0078
253. Zhen-Long, D., Li, X.-L., Jiao, L.-X., Shen, K.: Region duplication blind detection based on multiple feature combination. In: 2012 International Conference on Machine Learning and Cybernetics, Xi'an, China, pp. 17–21. IEEE (2012). https://doi.org/10.1109/ICMLC.2012.6358879
254. Rajkumar, R., Roy, S., Manglem Singh, K.: A robust and forensic transform for copy move digital image forgery detection based on dense depth block matching. Imaging Sci. J. **67**(6), 343–357 (2019). https://doi.org/10.1080/13682199.2019.1663069
255. Mahmood, T., Shah, M., Rashid, J., Saba, T., Nisar, M.W., Asif, M.: A passive technique for detecting copy-move forgeries by image feature matching. Multimed. Tools Appl. **79**(43–44), 31759–31782 (2020). https://doi.org/10.1007/s11042-020-09655-2

256. Wang, Y., Kang, X., Chen, Y.: Robust and accurate detection of image copy-move forgery using PCET-SVD and histogram of block similarity measures. J. Inf. Secur. Appl. **54** (2020). https://doi.org/10.1016/j.jisa.2020.102536
257. Meena, K.B., Tyagi, V.: A copy-move image forgery detection technique based on tetrolet transform. J. Inf. Secur. Appl. **52** (2020). https://doi.org/10.1016/j.jisa.2020.102481
258. Wu, S., Wu, Z., Huang, M.: Copy-move forgery detection via dimensionality reduction and double quantization feature. IEICE Trans. Inf. Syst. (2025). https://doi.org/10.1587/transinf.2024EDP7249
259. Shivakumar, B.L., Baboo, S.S.: Detection of region duplication forgery in digital images using SURF. Int. J. Comput. Sci. Issues. **8**(4), 199–205 (2011)
260. Wang, X., He, G., Tang, C., Han, Y., Wang, S.: Keypoints-based image passive forensics method for copy-move attacks. Int. J. Pattern Recognit. Artif. Intell. **30**(3) (2016). https://doi.org/10.1142/S0218001416550089
261. Soni, B., Das, P.K., Thounaojam, D.M.: Keypoints based enhanced multiple copy move forgeries detection system using density-based spatial clustering of application with noise clustering algorithm. IET Image Process. **12**(11), 2092–2099 (2018). https://doi.org/10.1049/iet-ipr.2018.5576
262. Wang, C., Zhang, Z., Li, Q., Zhou, X.: An image copy-move forgery detection method based on SURF and PCET. IEEE Access. **7**, 170032–170047 (2019). https://doi.org/10.1109/ACCESS.2019.2955308
263. Silva, E., Carvalho, T., Ferreira, A., Rocha, A.: Going deeper into copy-move forgery detection: exploring image telltales via multi-scale analysis and voting processes. J. Vis. Commun. Image Represent. **29**, 16–32 (2015). https://doi.org/10.1016/j.jvcir.2015.01.016
264. Yang, F., Li, J., Lu, W., Weng, J.: Copy-move forgery detection based on hybrid features. Eng. Appl. Artif. Intell. **59**, 73–83 (2017). https://doi.org/10.1016/j.engappai.2016.12.022
265. Yue, G., Duan, Q., Liu, R., Peng, W., Liao, Y., Liu, J.: SMDAF: a novel keypoint based method for copy-move forgery detection. IET Image Process. **16**(13), 3589–3602 (2022). https://doi.org/10.1049/ipr2.12578
266. Zhong, J.L., Pun, C.M.: Copy-move forgery detection using adaptive keypoint filtering and iterative region merging. Multimed. Tools Appl. **78**(18), 26313–26339 (2019). https://doi.org/10.1007/s11042-019-07817-5
267. Wang, J., et al.: Object-level copy-move forgery image detection based on inconsistency mining. In: WWW 2024 Companion - Companion Proceedings of the ACM Web Conference, pp. 943–946. Association for Computing Machinery (2024). https://doi.org/10.1145/3589335.3651540
268. Wang, J., et al.: Strong robust copy-move forgery detection network based on layer-by-layer decoupling refinement. Inf. Process. Manag. **61**(3) (2024). https://doi.org/10.1016/j.ipm.2024.103685
269. Liu, K., et al.: Copy move forgery detection based on keypoint and patch match. Multimed. Tools Appl. **78**(22), 31387–31413 (2019). https://doi.org/10.1007/s11042-019-07930-5
270. Zhao, K., et al.: CAMU-Net: copy-move forgery detection utilizing coordinate attention and multi-scale feature fusion-based up-sampling. Expert Syst. Appl. **238** (2024). https://doi.org/10.1016/j.eswa.2023.121918
271. Xiong, L., Xu, J., Yang, C.N., Zhang, X.: CMCF-Net: an end-to-end context multiscale cross-fusion network for robust copy-move forgery detection. IEEE Trans. Multimed. **26**, 6090–6101 (2024). https://doi.org/10.1109/TMM.2023.3345160
272. Vijayalakshmi K, N.V.S.K., Sasikala, J., Shanmuganathan, C.: Copy-paste forgery detection using deep learning with error level analysis. Multimed. Tools Appl. (2023). https://doi.org/10.1007/s11042-023-15594-5
273. Niu, P., Wang, C., Chen, W., Yang, H., Wang, X.: Fast and effective keypoint-based image copy-move forgery detection using complex-valued moment invariants. J. Vis. Commun. Image Represent. **77** (2021). https://doi.org/10.1016/j.jvcir.2021.103068

274. Weng, S., Zhu, T., Zhang, T., Zhang, C.: UCM-Net: a U-Net-like tampered-region-related framework for copy-move forgery detection. IEEE Trans. Multimed. **26**, 750–763 (2024). https://doi.org/10.1109/TMM.2023.3270629
275. Wang, X., Wang, X., Niu, P., Yang, H.: Accurate and robust image copy-move forgery detection using adaptive keypoints and FQGPCET-GLCM feature. Multimed. Tools Appl. **83**(1), 2203–2235 (2024). https://doi.org/10.1007/s11042-023-15499-3
276. Shi, Y., Weng, S., Yu, L., Li, L.: A copy-move forgery detection network based on selective sampling attention and low-cost two-step self-correlation calculation. IEEE Trans. Multimed. (2025). https://doi.org/10.1109/TMM.2025.3535369
277. Shi, Y., Weng, S., Yu, L., Li, L.: Lightweight and high-precision network for image copy-move forgery detection. IEEE Signal Process. Lett. **31**, 1409–1413 (2024). https://doi.org/10.1109/LSP.2024.3400055
278. Wu, Y., Abd-Almageed, W., Natarajan, P.: Image copy-move forgery detection via an end-to-end deep neural network. In: Proceedings - 2018 IEEE Winter Conference on Applications of Computer Vision, WACV 2018, pp. 1907–1915. IEEE (2018). https://doi.org/10.1109/WACV.2018.00211
279. Zhang, Y., Goh, J., Win, L.L., Thing, V.: Image region forgery detection: a deep learning approach. In: Cryptology and Information Security Series, pp. 1–11. IOS Press (2016). https://doi.org/10.3233/978-1-61499-617-0-1
280. Zhu, Y., Chen, C., Yan, G., Guo, Y., Dong, Y.: AR-Net: adaptive attention and residual refinement network for copy-move forgery detection. IEEE Trans. Industr. Inform. **16**(10), 6714–6723 (2020). https://doi.org/10.1109/TII.2020.2982705
281. Wu, Y., Abd-Almageed, W., Natarajan, P.: BusterNet: detecting copy-move image forgery with source/target localization. In: Computer Vision – ECCV 2018: 15th European Conference, Munich, Germany, September 8–14, 2018, Proceedings, Part VI, pp. 170–186. Springer-Verlag, Berlin, Heidelberg (2018). https://doi.org/10.1007/978-3-030-01231-1_11
282. Liu, B., Pun, C.-M.: Locating splicing forgery by adaptive-SVD noise estimation and vicinity noise descriptor. Neurocomputing. **387**, 172–187 (2020). https://doi.org/10.1016/j.neucom.2019.12.105
283. Liu, B., Pun, C.-M.: Locating splicing forgery by fully convolutional networks and conditional random field. Signal Process. Image Commun. **66**, 103–112 (2018). https://doi.org/10.1016/j.image.2018.04.011
284. Liu, B., Pun, C.-M.: Exposing splicing forgery in realistic scenes using deep fusion network. Inf. Sci. **526**, 133–150 (2020). https://doi.org/10.1016/j.ins.2020.03.099
285. Liu, B., Pun, C.M., Yuan, X.C.: Digital image forgery detection using JPEG features and local noise discrepancies. Sci. World J. (2014). https://doi.org/10.1155/2014/230425
286. Liu, Y., Guan, Q., Zhao, X., Cao, Y.: Image forgery localization based on multi-scale convolutional neural networks. In: Proceedings of the 6th ACM Workshop on Information Hiding and Multimedia Security (IH&MMSec '18), pp. 85–90. Association for Computing Machinery, New York, NY (2018). https://doi.org/10.1145/3206004.3206010
287. Mazumdar, A., Bora, P.K.: Estimation of lighting environment for exposing image splicing forgeries. Multimed. Tools Appl. **78**, 19839–19860 (2019). https://doi.org/10.1007/s11042-018-7147-2
288. Moghaddasi, Z., Jalab, H.A., Noor, R.M.: Image splicing forgery detection based on low-dimensional singular value decomposition of discrete cosine transform coefficients. Neural Comput. & Applic. **31**, 7867–7877 (2019). https://doi.org/10.1007/s00521-018-3586-y
289. Niyishaka, P., Bhagvati, C.: Image splicing detection technique based on Illumination-Reflectance model and LBP. Multimed. Tools Appl. **80**, 2161–2175 (2021). https://doi.org/10.1007/s11042-020-09707-7
290. Yildirim, E.O., Ulutaş, G.: Markov-based image splicing detection in the DCT high frequency region. In: 2018 International Conference on Artificial Intelligence and Data Processing (IDAP), Malatya, Turkey, pp. 1–4. IEEE (2018). https://doi.org/10.1109/IDAP.2018.8620870

291. Zhou, P., Han, X., Morariu, V.I., Davis, L.S.: Learning rich features for image manipulation detection. In: 2018 IEEE/CVF Conference on Computer Vision and Pattern Recognition (CVPR), Salt Lake City, UT, USA, pp. 1053–1061. IEEE (2018). https://doi.org/10.1109/CVPR.2018.00116
292. Pham, N.T., Lee, J.-W., Kwon, G.-R., Park, C.-S.: Efficient image splicing detection algorithm based on Markov features. Multimed. Tools Appl. **78**, 12405–12419 (2019). https://doi.org/10.1007/s11042-018-6792-9
293. Pun, C.-M., Liu, B., Yuan, X.-C.: Multi-scale noise estimation for image splicing forgery detection. J. Vis. Commun. Image Represent. **38**, 195–206 (2016). https://doi.org/10.1016/j.jvcir.2016.03.005
294. Rao, Y., Ni, J.: A deep learning approach to detection of splicing and copy-move forgeries in images. In: 2016 IEEE International Workshop on Information Forensics and Security (WIFS), Abu Dhabi, United Arab Emirates, pp. 1–6. IEEE (2016). https://doi.org/10.1109/WIFS.2016.7823911
295. Rao, Y., Ni, J., Zhao, H.: Deep learning local descriptor for image splicing detection and localization. IEEE Access. **8**, 25611–25625 (2020). https://doi.org/10.1109/ACCESS.2020.2970735
296. Pomari, T., Ruppert, G., Rezende, E., Rocha, A., Carvalho, T.: Image splicing detection through illumination inconsistencies and deep learning. In: 2018 25th IEEE International Conference on Image Processing (ICIP), Athens, Greece, pp. 3788–3792. IEEE (2018). https://doi.org/10.1109/ICIP.2018.8451227
297. Rhee, K.H.: Detection of spliced image forensics using texture analysis of median filter residual. IEEE Access. **8**, 103374–103384 (2020). https://doi.org/10.1109/ACCESS.2020.2999308
298. Salloum, R., Yuzhuo Ren, C.-C., Kuo, J.: Image splicing localization using a multi-task fully convolutional network (MFCN). J. Vis. Commun. Image Represent. **51**, 201–209 (2018). https://doi.org/10.1016/j.jvcir.2018.01.010
299. Shen, X., Shi, Z., Chen, H.: Splicing image forgery detection using textural features based on the grey level co-occurrence matrices. IET Image Process. **11**, 44–53 (2017). https://doi.org/10.1049/iet-ipr.2016.0238
300. Wang, J., Liu, R., Wang, H., Wu, B., Shi, Y.: Quaternion Markov splicing detection for color images based on quaternion discrete cosine transform. KSII Trans. Internet Inf. Syst. **14**, 2981–2996 (2020). https://doi.org/10.3837/tiis.2020.07.014
301. Wang, J., Ni, Q., Liu, G., Luo, X., Jha, S.K.: Image splicing detection based on convolutional neural network with weight combination strategy. J. Inf. Secur. Appl. **54**, 102523 (2020). https://doi.org/10.1016/j.jisa.2020.102523
302. Wang, R., Lu, W., Li, J., Xiang, S., Zhao, X., Wang, J.: Digital image splicing detection based on Markov features in QDCT and QWT domain. Int. J. Digit. Crime For. **10**(4), 90–107 (2018). https://doi.org/10.4018/IJDCF.2018100107
303. Wang, X., Zhang, Q., Jiang, C., Zhang, Y.: Coarse-to-fine grained image splicing localization method based on noise level inconsistency. In: 2020 International Conference on Computing, Networking and Communications (ICNC), Big Island, HI, USA, pp. 79–83. IEEE (2020). https://doi.org/10.1109/ICNC47757.2020.9049720
304. Yang, W., Wang, Z., Xiao, B., Liu, X., Zheng, Y., Ma, J.: Controlling neural learning network with multiple scales for image splicing forgery detection. ACM Trans. Multimedia Comput. Commun. Appl. **16**(4), Article 124 (2020). https://doi.org/10.1145/3408299
305. Wu, J., Chang, X., Yang, T., Feng, K.: Blind forensic method based on convolutional neural networks for image splicing detection. In: 2019 IEEE 5th International Conference on Computer and Communications (ICCC), Chengdu, China, pp. 2014–2018. IEEE (2019). https://doi.org/10.1109/ICCC47050.2019.9064258
306. Wu, Y., Abdalmageed, W., Natarajan, P.: ManTra-Net: manipulation tracing network for detection and localization of image forgeries with anomalous features. In: 2019 IEEE/CVF Conference on Computer Vision and Pattern Recognition (CVPR), Long Beach, CA, USA, pp. 9535–9544. IEEE (2019). https://doi.org/10.1109/CVPR.2019.00977

307. Yao, H., Xu, M., Qiao, T., Wu, Y., Zheng, N.: Image forgery detection and localization via a reliability fusion map. Sensors (Switzerland). **20**, 1–18 (2020). https://doi.org/10.3390/s20226668
308. Ye, K., Dong, J., Wang, W., Peng, B., Tan, T.: Feature pyramid deep matching and localization network for image forensics. In: 2018 Asia-Pacific Signal and Information Processing Association Annual Summit and Conference (APSIPA ASC), Honolulu, HI, USA, pp. 1796–1802. IEEE (2018). https://doi.org/10.23919/APSIPA.2018.8659464
309. Yildirim, E.O., Ulutaş, G.: Image splicing detection with dwt domain extended Markov features. In: 2018 26th Signal Processing and Communications Applications Conference (SIU), Izmir, Turkey, pp. 1–4. IEEE (2018). https://doi.org/10.1109/SIU.2018.8404325
310. Yıldırım, E.O., Ulutaş, G.: Augmented features to detect image splicing on SWT domain. Expert Syst. Appl. **131**, 81–93 (2019). https://doi.org/10.1016/j.eswa.2019.04.036
311. Zampoglou, M., Papadopoulos, S., Kompatsiaris, Y.: Large-scale evaluation of splicing localization algorithms for web images. Multimed. Tools Appl. **76**, 4801–4834 (2017). https://doi.org/10.1007/s11042-016-3795-2
312. Zampoglou, Z.M., Papadopoulos, S., Kompatsiaris, Y.: Detecting image splicing in the wild (WEB). In: 2015 IEEE International Conference on Multimedia & Expo Workshops (ICMEW), Turin, Italy, pp. 1–6. IEEE (2015). https://doi.org/10.1109/ICMEW.2015.7169839
313. Zeng, H., Peng, A., Lin, X.: Exposing image splicing with inconsistent sensor noise levels. Multimed. Tools Appl. **79**, 26139–26154 (2020). https://doi.org/10.1007/s11042-020-09280-z
314. Zeng, H., Zhan, Y., Kang, X., et al.: Image splicing localization using PCA-based noise level estimation. Multimed. Tools Appl. **76**, 4783–4799 (2017). https://doi.org/10.1007/s11042-016-3712-8
315. Zhang, Q., Lu, W., Wang, R., et al.: Digital image splicing detection based on Markov features in block DWT domain. Multimed. Tools Appl. **77**, 31239–31260 (2018). https://doi.org/10.1007/s11042-018-6230-z
316. Zhang, Y., Zhang, J., Xu, S.: A hybrid convolutional architecture for accurate image manipulation localization at the pixel-level. Multimed. Tools Appl. **80**, 23377–23392 (2021). https://doi.org/10.1007/s11042-020-10211-1
317. Zhang, Z., Zhang, Y., Zhou, Z., Luo, J.: Boundary-based image forgery detection by fast shallow CNN. In: 2018 24th International Conference on Pattern Recognition (ICPR), Beijing, China, pp. 2658–2663. IEEE (2018). https://doi.org/10.1109/ICPR.2018.8545074
318. Zhu, N., Li, Z.: Blind image splicing detection via noise level function. Signal Process. Image Commun. **68**, 181–192 (2018). https://doi.org/10.1016/j.image.2018.07.012
319. Meena, K.B., Tyagi, V.: Image forgery detection: survey and future directions. Data Eng. Appl. **2**, 163–194 (2019). https://doi.org/10.1007/978-981-13-6351-1_14
320. Meena, K.B., Tyagi, V.: A deep learning based method for image splicing detection. J. Phys. Conf. Ser. **1714**, 012038 (2021). https://doi.org/10.1088/1742-6596/1714/1/012038
321. Meena, K.B., Tyagi, V.: Image splicing forgery detection techniques: a review. In: Advances in Computing and Data Sciences. ICACDS 2021. Communications in Computer and Information Science, vol. 1441. Springer, Cham (2021). https://doi.org/10.1007/978-3-030-88244-0_35
322. Meena, K.B., Tyagi, V.: Image splicing forgery detection using noise level estimation. Multimed. Tools Appl. **82**, 13181–13198 (2023). https://doi.org/10.1007/s11042-021-11483-x
323. Liu, G., Reda, F., Shih, K., Wang, T.C., Tao, A., Catanzaro, B.: Image inpainting for irregular holes using partial convolutions. (2018). https://doi.org/10.48550/arXiv.1804.07723
324. Majumder, M.T.H., Alim Al Islam, A.B.M.: A tale of a deep learning approach to image forgery detection. In: 2018 5th international conference on networking, systems and security (NSysS), pp. 1–9. IEEE (2018). https://doi.org/10.1109/NSysS.2018.8631389
325. Marra, F., Gragnaniello, D., Verdoliva, L., Poggi, G.: A full-image full-resolution end-to-end-trainable CNN framework for image forgery detection. IEEE Access. **8** (2020). https://doi.org/10.1109/ACCESS.2020.3009877
326. Muzaffer, G., Ulutas, G.: A new deep learning-based method to detection of copy-move forgery in digital images. In: 2019 Scientific meeting on electrical-electronics biomedical

engineering and computer science (EBBT), pp. 1–4. IEEE (2019). https://doi.org/10.1109/EBBT.2019.8741657
327. Nguyen, H.H., Yamagishi, J., Echizen, I.: Capsule-forensics: using capsule networks to detect forged images and videos. In: ICASSP 2019 - 2019 IEEE International Conference on Acoustics, Speech and Signal Processing (ICASSP), Brighton, UK, pp. 2307–2311. IEEE (2019). https://doi.org/10.1109/ICASSP.2019.8682602
328. Nightingale, S.J., Wade, K.A., Watson, D.G.: Can people identify original and manipulated photos of real-world scenes? Cogn. Res. Princ. Implic. **2**(1), 1–21 (2017). https://doi.org/10.1186/s41235-017-0067-2
329. Ouyang, J., Liu, Y., Liao, M.: Copy-move forgery detection based on deep learning. In: 2017 10th International Congress on Image and Signal Processing, BioMedical Engineering and Informatics (CISP-BMEI), pp. 1–5. IEEE (2017). https://doi.org/10.1109/CISP-BMEI.2017.8301940
330. Piva, A.: An overview on image forensics. Int. Sch. Res. Not. **2013**, 496701 (2013). https://doi.org/10.1155/2013/496701
331. Popescu, A.C., Farid, H.: Exposing digital forgeries by detecting duplicated image regions. Tech. Rep. TR2004-515 (2004)
332. Popescu, A.C., Farid, H.: Exposing digital forgeries by detecting traces of re-sampling. IEEE Trans. Signal Process. **53**(2), 758–767 (2005). https://doi.org/10.1109/TSP.2004.839932
333. Qureshi, M.A., Deriche, M.: A bibliography of pixel-based blind image forgery detection techniques. Signal Process. Image Commun. **39**, 46–74 (2015). https://doi.org/10.1016/j.image.2015.08.008
334. Rössler, A., Cozzolino, D., Verdoliva, L., Riess, C., Thies, J., Nießner, M.: Faceforensics++: learning to detect manipulated facial images. (2019). https://doi.org/10.48550/arXiv.1901.08971
335. Roy, S., Sun, Q.: Robust hash for detecting and localizing image tampering. In: 2007 IEEE International Conference on Image Processing, pp. VI-117–VI-120. IEEE (2007). https://doi.org/10.1109/ICIP.2007.4379535
336. Shen, C., Kasra, M., Pan, P., Bassett, G.A., Malloch, Y., O'Brien, J.F.: Fake images: the effects of source, intermediary, and digital media literacy on contextual assessment of image credibility online. New Media Soc. **21**(2), 438–463 (2019). https://doi.org/10.1177/1461444818799526
337. Spohr, D.: Fake news and ideological polarization: filter bubbles and selective exposure on social media. Bus. Inf. Rev. **34**(3), 150–160 (2017). https://doi.org/10.1177/0266382117722446
338. Thakur, R., Rohilla, R.: Copy-move forgery detection using residuals and convolutional neural network framework: a novel approach. In: 2019 2nd International Conference on Power Energy, Environment and Intelligent Control PEEIC, pp. 561–564. IEEE (2019). https://doi.org/10.1109/PEEIC47157.2019.8976868
339. Reddy, B.S., Lavanya, K., Vamsi, K., Seshi Reddy, K.S., Guru Lingaraju, M.: A robust approach for handwritten signature verification through deep learning. In: 2025 7th International Conference on Intelligent Sustainable Systems (ICISS), India, pp. 1386–1393. IEEE (2025). https://doi.org/10.1109/ICISS63372.2025.11076424
340. Liu, C., Zhang, G., Guo, S., Li, Q., Jeon, G., Gao, M.: Context-aware deepfake detection for securing AI-driven financial transactions. IEEE Trans. Comput. Soc. Syst. https://doi.org/10.1109/TCSS.2025.3577753
341. Narale, D.B., Pujar, A.M., Pardeshi, K.R.: Optimizing currency classification and counterfeit detection with machine learning algorithms. In: 2025 3rd International Conference on Inventive Computing and Informatics (ICICI), Bangalore, India, pp. 1041–1046. IEEE (2025). https://doi.org/10.1109/ICICI65870.2025.11069825
342. Wu, H., Chen, Y., Zhou, J., Li, Y.: Rethinking image forgery detection via soft contrastive learning and unsupervised clustering. IEEE Trans. Dependable Secure Comput. https://doi.org/10.1109/TDSC.2025.3583167

343. Wang, J., Pan, G., Sun, D., Li, J., Zhang, J.: AFAN: an attention-driven forgery adversarial network for blind image inpainting. IEEE Trans. Multimed. https://doi.org/10.1109/TMM.2025.3590914
344. Xu, J.: EdgeAdapter for image forgery localization. In: 2025 6th International Conference on Computer Vision, Image and Deep Learning (CVIDL), Ningbo, China, pp. 1156–1160. IEEE (2025). https://doi.org/10.1109/CVIDL65390.2025.11085809
345. Magoo, K., Amity, T.C., Gandhi, R., Sharma, M.: Deepfake image and forged signature detection using machine learning. In: 2025 International Conference on Engineering, Technology & Management (ICETM), Oakdale, NY, USA, pp. 1–6. IEEE (2025). https://doi.org/10.1109/ICETM63734.2025.11051925
346. Liu, L., Sun, P., Lang, Y., Li, J.: CFA-based splicing forgery localization method via statistical analysis. IET Signal Process. (2024). https://doi.org/10.1049/2024/9929900
347. Özden, M., Şahin, C.: A comparative study for localization of forgery regions in images. IEEE Access. https://doi.org/10.1109/ACCESS.2025.3591571
348. Vasudevan, M., Sireesha, C., Sumiya, N., Vardhan, K.V., Sagar, T.: CNN based error level analysis (ELA) with scores to detect digital image manipulation. In: 2025 3rd International Conference on Self Sustainable Artificial Intelligence Systems (ICSSAS), Erode, India, pp. 1541–1547. IEEE (2025). https://doi.org/10.1109/ICSSAS66150.2025.11081309
349. Verdoliva, L.: Media forensics and deepfakes: an overview. IEEE J. Sel. Top. Signal Process. (2020). https://doi.org/10.1109/JSTSP.2020.3002101
350. Lv, X., Su, J., Gao, Z., Zhang, L.: Improved counterfeit image detection for YOLOv8. In: 2025 5th International Conference on Neural Networks, Information and Communication Engineering (NNICE), Guangzhou, China, pp. 452–456. IEEE (2025). https://doi.org/10.1109/NNICE64954.2025.11064652
351. Wang, Y., Fan, L., Cui, J., Zheng, H.: A generalization-enhanced method for forged face detection based on high-frequency features. IEICE Trans. Commun. https://doi.org/10.23919/transcom.2024EBP3212
352. Yang, Y., Deng, Y., Yi, Q.: Self-attention forged images detection network with three-branch multi-feature fusion. In: 2025 IEEE 14th Data Driven Control and Learning Systems (DDCLS), Wuxi, China, pp. 2253–2258. IEEE (2025). https://doi.org/10.1109/DDCLS66240.2025.11064938
353. Nguyen, A.D., Kim, H.-Y., Nguyen, H.N.: TALIU: a novel decoder and augmentation strategy for boosting tampered document image detection. IEEE Access. **13**, 70340–70351 (2025). https://doi.org/10.1109/ACCESS.2025.3560360
354. Phan-Ho, A.-T., Retraint, F.: A comparative study of Bayesian and Dempster-Shafer fusion on image forgery detection. IEEE Access. **10**, 99268–99281 (2022). https://doi.org/10.1109/ACCESS.2022.3206543
355. Zakey, A., Bawantha, D., Shehara, D., Hasara, N., Abeywardena, K.Y., Fernando, H.: A dual-branch CNN and metadata analysis approach for robust image tampering detection. In: 2025 13th International Symposium on Digital Forensics and Security (ISDFS), Boston, MA, USA, pp. 1–6. IEEE (2025). https://doi.org/10.1109/ISDFS65363.2025.11012015
356. Peng, C., Luo, X., Liu, D., Wang, N., Hu, R., Gao, X.: Semantic token transformer for face forgery detection. IEEE Trans. Inf. Forensics Secur. **20**, 4904–4914 (2025). https://doi.org/10.1109/TIFS.2025.3567110
357. Zhang, G., Li, Q., Gao, M., Guo, S., Jeon, G., Abdelmoniem, A.M.: Space–frequency and global–local attentive networks for sequential deepfake detection. IEEE Trans. Comput. Soc. Syst. https://doi.org/10.1109/TCSS.2025.3541346
358. Shao, H.-C., Liao, Y.-R., Tseng, T.-Y., Chuo, Y.-L., Lin, F.-Y.: Copy-move detection in optical microscopy: a segmentation network and a dataset. IEEE Signal Process. Lett. **32**, 1106–1110 (2025). https://doi.org/10.1109/LSP.2025.3547273
359. Wang, H., Deng, C., Zhao, Z.: Knowledge-guided prompt learning for deepfake facial image detection. In: ICASSP 2025 - 2025 IEEE International Conference on Acoustics, Speech

and Signal Processing (ICASSP), Hyderabad, India, pp. 1–5. IEEE (2025). https://doi.org/10.1109/ICASSP49660.2025.10889149
360. Reddy, J.K., Aman, D., Yashwanth, B., Sowmya, P., Srikanth, R., Vasantha, S.: Enhancing web security with VulnGuard: an advanced vulnerability scanning approach. In: 2025 International Conference on Engineering, Technology & Management (ICETM), Oakdale, NY, USA, pp. 1–7. IEEE (2025). https://doi.org/10.1109/ICETM63734.2025.11051982
361. Rani, J., Anand, A., Shivani, S.: Denoising CNN assisted ECG signal watermarking for secure transmission in e-healthcare applications. IEEE Multimed. https://doi.org/10.1109/MMUL.2025.3583001
362. Wang, J., et al.: Copy-move forgery image detection based on cross-scale modeling and alternating refinement. IEEE Trans. Multimed. https://doi.org/10.1109/TMM.2025.3543057
363. Wang, J., et al.: Fighting malicious media data: a survey on tampering detection and deepfake detection. Proc. IEEE. https://doi.org/10.1109/JPROC.2025.3576367
364. Wu, K., Li, L., Li, Q.: AGU2-net: multi-scale U2-net enhanced by attention gate mechanism for image tampering localization. IEEE Access. **13**, 99659–99671 (2025). https://doi.org/10.1109/ACCESS.2025.3577221
365. Sowmya, M., Suriya, S.S.R., Banu, D.F., Kaleeswari, K., Gayathri, L., Jothika, K.: Image copyright protection using quantum watermarking. In: 2025 3rd International Conference on Artificial Intelligence and Machine Learning Applications Theme: Healthcare and Internet of Things (AIMLA), Namakkal, India, pp. 1–11. IEEE (2025). https://doi.org/10.1109/AIMLA63829.2025.11040769
366. Zhang, L., et al.: Rethinking image forgery detection and localization via regression perspective. IEEE Trans. Emerg. Top. Comput. Intell. https://doi.org/10.1109/TETCI.2025.3543837
367. Zhuo, L., Luo, S., Tan, S., Chen, H., Li, B., Huang, J.: Evading detection actively: toward anti-forensics against forgery localization. IEEE Trans. Dependable Secure Comput. **22**(2), 852–869 (2025). https://doi.org/10.1109/TDSC.2025.3528062
368. Mao, M., Yan, C., Wang, J., Yang, J.: Leveraging pixel difference feature for Deepfake detection. IEEE Trans. Emerg. Top. Comput. Intell. **9**(4), 3178–3188 (2025). https://doi.org/10.1109/TETCI.2025.3548803
369. Yang, M., Qi, B., Ma, R., Xian, Y., Ma, B.: HashShield: a robust deepfake forensic framework with separable perceptual hashing. IEEE Signal Process. Lett. **32**, 1186–1190 (2025). https://doi.org/10.1109/LSP.2025.3547664
370. Xiao, Q., Wu, Y., Tao, S.: Optical diffraction field-based palmprint recognition. IEEE Sensors J. **25**(14), 27230–27237 (2025). https://doi.org/10.1109/JSEN.2025.3577623
371. Zhang, Q., Yang, G., Li, Z., Lu, K.: DSBU-Net: dual-stream branch union network for Image manipulation localization. In: 2025 4th International Symposium on Computer Applications and Information Technology (ISCAIT), Xi'an, China, pp. 696–704. IEEE (2025). https://doi.org/10.1109/ISCAIT64916.2025.11010267
372. Sun, R., Yu, X., Wang, F., Da, Z., Zhang, Y., Gao, J.: Frequency-assisted temporal upsampling artifacts representation learning for face forgery detection. IEEE Trans. Biom. Behav. Identity Sci. https://doi.org/10.1109/TBIOM.2025.3569645
373. Vinolin, V., Sucharitha, M.: Hierarchical categorization and review of recent techniques on image forgery detection. Comput. J. **64**(11), 1692–1704 (2019). https://doi.org/10.1093/comjnl/bxz148
374. Lu, X., Cheng, F., Luo, G., Zhu, Y.: MFDF-IML: multi-feature dynamic fusion for image manipulation localization. In: 2025 28th International Conference on Computer Supported Cooperative Work in Design (CSCWD), Compiegne, France, pp. 686–691. IEEE (2025). https://doi.org/10.1109/CSCWD64889.2025.11033390
375. Qiu, X., Shi, C., Li, X., Wu, M., Abdullahi, S.M., Liu, Y.: S-faster R-CNN: intraspectral similarity learning for audio copy-move forgery localization in IoT security. IEEE Internet Things J. **12**(15), 30185–30202 (2025). https://doi.org/10.1109/JIOT.2025.3569678

376. Wei, Y., Liu, H., Yuan, X., Bi, X., Xiao, B.: Let images speak more: an efficient method for detecting image manipulation history. IEEE Trans. Circuits Syst. Video Technol. https://doi.org/10.1109/TCSVT.2025.3571767
377. Wu, Y., et al.: SONICUMOS: an enhanced active face liveness detection system via ultrasonic and video signals. IEEE Trans. Mob. Comput. https://doi.org/10.1109/TMC.2025.3565689
378. Zhang, Y., Wang, T., Yu, Z., Gao, Z., Shen, L., Chen, S.: MFCLIP: multi-modal fine-grained CLIP for generalizable diffusion face forgery detection. IEEE Trans. Inf. Forensics Secur. **20**, 5888–5903 (2025). https://doi.org/10.1109/TIFS.2025.3576577
379. Shi, Z., Chen, H., Jia, Y., Zhang, D., Lu, W., Yang, X.: Customized transformer adapter with frequency masking for deepfake detection. IEEE Trans. Inf. Forensics Secur. **20**, 5904–5918 (2025). https://doi.org/10.1109/TIFS.2025.3574983
380. Sun, Z., Ruan, N., Li, J.: DDL: effective and comprehensible interpretation framework for diverse deepfake detectors. IEEE Trans. Inf. Forensics Secur. **20**, 3601–3615 (2025). https://doi.org/10.1109/TIFS.2025.3553803
381. Zhu, Z., Li, J., Wen, Y.: Self-optimization training for weakly supervised image manipulation localization. In: ICASSP 2025 - 2025 IEEE International Conference on Acoustics, Speech and Signal Processing (ICASSP), Hyderabad, India, pp. 1–5. IEEE (2025). https://doi.org/10.1109/ICASSP49660.2025.10889843

Made in the USA
Monee, IL
03 May 2026

49438516R00063